BIG IDEAS MATH®
Advanced 1

AF869762

Resources by Chapter

- Family and Community Involvement
- Start Thinking! and Warm Up
- Extra Practice (A and B)
- Enrichment and Extension
- Puzzle Time
- Technology Connection

Assessment Book

- Quizzes
- Chapter Tests
- Standards Assessment with Item Analysis
- Alternative Assessments
- End-of-Course Tests

BIG IDEAS LEARNING®

Erie, Pennsylvania

3 0099 03312290 0

Copyright © by Big Ideas Learning, LLC. All rights reserved.

Permission is hereby granted to teachers to reprint or photocopy in classroom quantities only the pages or sheets in this work that carry a Big Ideas Learning copyright notice, provided each copy made shows the copyright notice. These pages are designed to be reproduced by teachers for use in their classes with accompanying Big Ideas Learning material, provided each copy made shows the copyright notice. Such copies may not be sold and further distribution is expressly prohibited. Except as authorized above, prior written permission must be obtained from Big Ideas Learning, LLC to reproduce or transmit this work or portions thereof in any other form or by any other electronic or mechanical means, including but not limited to photocopying and recording, or by any information storage or retrieval system, unless expressly permitted by copyright law. Address inquiries to Permissions, Big Ideas Learning, LLC, 1762 Norcross Road, Erie, PA 16510.

Big Ideas Learning and *Big Ideas Math* are registered trademarks of Larson Texts, Inc.

ISBN 13: 978-1-60840-577-0
ISBN 10: 1-60840-577-X

23456789-VLP-17 16 15 14

Contents

Resources by Chapter

Assessment Book

Copyright © Big Ideas Learning, LLC
All rights reserved.

Chapter 11

Copyright © Big Ideas Learning, LLC
All rights reserved.

Name____ Date__________

Chapter 11 Integers

Dear Family,

Hiking can be good exercise and a nice opportunity to talk with your family. Whether you walk to a nearby park, or travel to a favorite hiking trail, you usually plan to end up where you start. How far have you traveled on your walk, then? In a sense, you have gone nowhere—you have traveled zero distance.

This isn't the whole story, though. What also happens is that you travel a certain distance there and then the same distance back. The two distances are in opposite directions, so they bring you back to your starting point. But if you want to know how far you have walked, you talk about the distance without regard to direction—the *absolute value*. If the park is a mile away, you walk one mile there and one mile back—two miles in total.

The same reasoning applies when you walk up a hill. If you climb a 300-foot hill, you are going up 300 feet. If you want to end up back at the bottom, you must eventually climb down 300 feet. How much have you climbed in total?

While you are walking, keep track of how far you have traveled. You might talk with your student about the following:

- When is it helpful to assign direction (positive or negative) to each part of the walk?
- When is it helpful to ignore direction and just use the absolute value of the distance?

Usually we express the up direction as a positive number and the down direction as a negative number. Talk with your student about why that might be the case. With your student, think of situations where the reverse might be more convenient.

Enjoy the sunshine while you walk and talk with your student!

Copyright © Big Ideas Learning, LLC
All rights reserved.

Nombre ____ Fecha __________

Capítulo 11 Números enteros

Estimada Familia:

Salir de excursión puede ser un buen ejercicio y una buena oportunidad para conversar con su familia. Ya sea que caminen a un parque cercano o viajen a su sendero de excursión favorito, generalmente planean terminar donde empezaron. Entonces, ¿cuánto han recorrido en su caminata? En cierto sentido, no han ido a ningún lugar—han recorrido una distancia cero.

Sin embargo, aquí no acaba todo. Lo que también sucede es que recorrieron una cierta distancia hacia allá y luego recorrieron la misma distancia de regreso. Las dos distancias están en direcciones opuestas, así que los traen de vuelta a su punto de partida. Pero si desean saber cuánto han caminado, se habla sobre la distancia sin considerar la dirección—es decir, el *valor absoluto*. Si el parque queda a una milla de distancia, ustedes caminan una milla hacia allá y una milla de regreso—dos millas, en total.

El mismo razonamiento se aplica cuando se trepa por una colina. Si trepa una colina de 300 pies, está ascendiendo 300 pies. Si desea terminar de regreso en la parte inferior, eventualmente tendrá que bajar 300 pies. ¿Cuánto ha trepado en total?

Mientras está caminando, lleve un registro de la distancia que ha recorrido. Querrá hablar con su estudiante acerca de lo siguiente:

- ¿Cuándo es útil asignar una dirección (positiva o negativa) a cada parte de la caminata?
- ¿Cuándo es útil ignorar una dirección y sólo usar el valor absoluto de la distancia?

Generalmente expresamos la dirección ascendente con un número positivo y la descendente con un número negativo. Converse con su estudiante acerca de por qué sería ese el caso. Con su estudiante, piense en situaciones en donde lo inverso podría ser más conveniente.

¡Disfrute de la luz solar mientras camina y conversa con su estudiante!

Copyright © Big Ideas Learning, LLC
All rights reserved.

Activity 11.1 Start Thinking!

For use before Activity 11.1

Explain how football is like solving a math problem.

Activity 11.1 Warm Up

For use before Activity 11.1

Copy and complete the statement using < or >.

1. 12 __?__ 14

2. 36 __?__ 26

3. −2 __?__ −5

4. −15 __?__ −8

5. 13 __?__ −10

6. −20 __?__ 19

Copyright © Big Ideas Learning, LLC
All rights reserved.

Lesson 11.1 Start Thinking!

For use before Lesson 11.1

When you go to school in the morning, you travel in one direction. Then returning home, you travel in the other direction. How does this compare to a number line where one direction is positive and the other is negative? Do you ever travel in a negative direction?

Lesson 11.1 Warm Up

For use before Lesson 11.1

Find the absolute value.

1. $|15|$ **2.** $|-23|$ **3.** $|7|$ **4.** $|-35|$

5. $|-43|$ **6.** $|0|$ **7.** $|39|$ **8.** $|-212|$

Copyright © Big Ideas Learning, LLC
All rights reserved.

Name ______________________________ Date __________

11.1 Practice A

Find the absolute value.

1. $|-7|$
2. $|12|$
3. $|-13|$
4. $|0|$

Copy and complete the statement using <, >, or =.

5. $|-4|$ __?__ 2
6. 7 __?__ $|-7|$
7. $|8|$ __?__ 5

8. While playing a game, you move back 5 spaces with your roll of the number cube. Your friend moves forward 3 spaces. Write each amount as an integer.

Order the values from least to greatest.

9. $-1, |5|, |4|, 8, |-1|$
10. $|2|, 0, |5|, 6, |3|$

Simplify the expression.

11. $|-19|$
12. $-|-8|$
13. $-|13|$

14. You are kite sailing on the ocean. The table gives your height at different times.

Time (seconds)	0	1	2	3
Height (feet)	2	4	6	8

  a. How many feet do you move each second?

  b. What is your speed? Give the units.

  c. Is your velocity positive or negative?

  d. What is your velocity? Give the units.

15. Use a number line.

  a. Graph and label the following points on a number line: $T = 1$, $L = -8$, $E = 4$, $A = -5$. What word do the letters spell?

  b. Graph and label the absolute value of each point in part (a). What word do the letters spell now?

16. Write an integer whose absolute value is greater than itself.

Copyright © Big Ideas Learning, LLC
All rights reserved.

Name ____ Date __________

11.1 Practice B

Copy and complete the statement using $<$, $>$, or $=$.

1. $|-23| \ \underline{\ ?\ } \ 23$
2. $|-142| \ \underline{\ ?\ } \ |-157|$
3. $-|-78| \ \underline{\ ?\ } \ 52$

4. You and your friend are swimming against the current. You move forward 15 feet. Your friend is not a strong swimmer, so he moves back 6 feet. Write each amount as an integer.

Order the values from least to greatest.

5. $14, |-25|, -|-34|, 28, |0|$
6. $|-16|, 10, |25|, -16, |-43|$

Simplify the expression.

7. $|-249|$
8. $-|183|$
9. $-|-153|$

10. The boiling point of a liquid is the temperature at which the vapor pressure of the liquid equals the environmental pressure surrounding the liquid.

Substance	Hydrogen	Oxygen	Iodine	Phosphorus
Boiling Point (°C)	−253	−183	184	280

   a. Which substance in the table has the highest boiling point?

   b. Is the boiling point of oxygen or iodine closer to 0°C?

11. You are riding on a rollercoaster.

   a. Your velocity is 13 feet per second. Are you moving up or moving down?

   b. What is your speed in part (a)? Give the units.

   c. Your velocity is -17 feet per second. Are you moving up or moving down?

   d. What is your speed in part (c)? Give the units.

12. There is one integer for which there does not exist another integer with the same absolute value. What is that integer?

Determine whether the statement is *true* or *false*. Explain your reasoning.

13. The absolute value of 3 above par is the same as the absolute value of 3 below par.

14. If $x < 0$, then $|x| < x$.

Copyright © Big Ideas Learning, LLC
All rights reserved.

Name____ Date__________

11.1 Enrichment and Extension

Reasoning with Integers

Assume $a > 0$ and $b < 0$. Determine whether the statement is *always, sometimes,* or *never* true.

1. $a > b$
2. $a < b$
3. $|a| > |b|$
4. $|a| < |b|$
5. $-a > -b$
6. $-a < -b$
7. $-|a| < -|b|$
8. $|-a| < |-b|$

Let n be an integer. Determine whether the possible values of n are *all integers, all positive integers, all negative integers, all positive integers and zero, all negative integers and zero,* or *none.*

9. $n < 0$
10. $n > 0$
11. $|n| < 0$
12. $|n| \geq 0$
13. $|-n| < 0$
14. $|-n| \geq 0$
15. $-|n| \leq 0$
16. $-|n| > 0$
17. $-|-n| \leq 0$
18. $-|-n| > 0$
19. $|n| = -n$
20. $-|n| = n$
21. $|n| = n$
22. $|-n| = n$
23. $-|-n| = -n$
24. $-|-n| = n$
25. $|n| < n$
26. $|n| > n$
27. $|n| > -n$
28. $-|n| < n$
29. $|n| > |-n|$
30. $|n| \geq -|n|$
31. $-|-n| < n$
32. $-|-n| \leq |n|$
33. $|n| + |n| < 2n$
34. $|n| + |n| = 2|n|$
35. $-|n| + |n| = 0$
36. $-|-n| + |-n| = 0$

For each exercise number, use your answers and the key below to color the cell. Do not color the cells that have a zero in them.

Always = Orange

Sometimes = Green

Never = Brown

All Integers = Brown

All Positive Integers = Brown

All Negative Integers = Green

All Positive Integers and Zero = Green

All Negative Integers and Zero = Brown

None = Brown

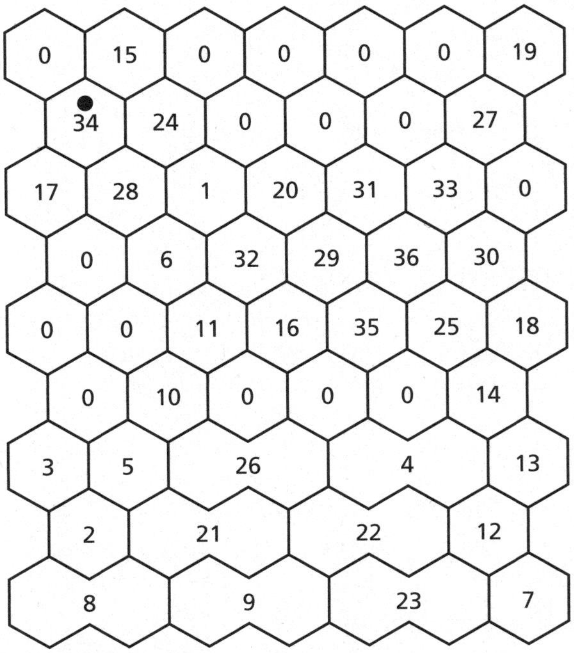

©Buki Toys Ltd.

Copyright © Big Ideas Learning, LLC
All rights reserved.

Name ______________________________ Date __________

11.1 Puzzle Time

What Can You Serve, But Never Eat?

Write the letter of each answer in the box containing the exercise number.

Find the absolute value.

1. $|12|$
2. $|-9|$
3. $|20|$
4. $|-10|$

Complete the statement using <, >, or =.

5. $3 \underline{\ ?\ } |-8|$
6. $4 \underline{\ ?\ } |-4|$
7. $|-6| \underline{\ ?\ } -6$

Simplify the expression.

8. $-|-13|$
9. $|-55|$
10. $-|2|$
11. A fishfinder is an instrument on a boat that indicates where fish are located. Are the fish closest to the surface of the water at -20 feet or -30 feet?

Answers

S. 10	**B.** -13
E. -2	**T.** 20
G. -55	**A.** 9
I. 12	**F.** 2
S. -20	**B.** -9
H. -12	**D.** 13
L. 55	**R.** -10
A. $<$	**N.** $>$
L. $=$	**N.** -20 feet
P. -30 feet	

5		3	10	11	7	1	4		8	2	9	6

Copyright © Big Ideas Learning, LLC
All rights reserved.

Activity 11.2 Start Thinking!

For use before Activity 11.2

Explain how playing a round of golf is like solving an integer problem.

Activity 11.2 Warm Up

For use before Activity 11.2

Add.

1. $10 + 12$

2. $14 + 28$

3. $26 + 32 + 19$

4. $47 + 35 + 68$

5. $12 + 33 + 59 + 18$

6. $82 + 13 + 29 + 97$

Copyright © Big Ideas Learning, LLC
All rights reserved.

Lesson 11.2 Start Thinking!

For use before Lesson 11.2

The temperature first rises 10 degrees and then falls 12 degrees. Is the end temperature greater than or less than the starting temperature? How does this compare to adding integers?

Lesson 11.2 Warm Up

For use before Lesson 11.2

Add.

1. $9 + 4$

2. $-5 + (-1)$

3. $-8 + 13$

4. $12 + (-8)$

5. $-10 + 6$

6. $-14 + 14$

Copyright © Big Ideas Learning, LLC
All rights reserved.

Name________________________ Date__________

11.2 Practice A

Add.

1. $8 + 2$ **2.** $-5 + (-3)$ **3.** $-9 + (-3)$ **4.** $6 + (-6)$

5. $4 + (-4)$ **6.** $9 + (-6)$ **7.** $5 + (-2)$ **8.** $7 + (-13)$

9. $-18 + 1$ **10.** $-12 + (-5)$ **11.** $0 + (-7)$ **12.** $12 + (-15)$

13. Your bank account has a balance of $-\$21$. You deposit $\$50$. What is your new balance?

Tell how the Commutative and Associative Properties of Addition can help you find the sum mentally. Then find the sum.

14. $8 + (-5) + (-8)$ **15.** $-4 + 9 + 4$ **16.** $-5 + 12 + (-7)$

Add.

17. $7 + 5 + (-2)$ **18.** $-13 + 7 + (-3)$ **19.** $17 + (-5) + (-1)$

20. $4 + 8 + (-8)$ **21.** $-12 + (-4) + 9$ **22.** $-10 + 10 + (-3)$

23. $(-11) + 5 + (-12)$ **24.** $7 + 15 + (-7)$ **25.** $-12 + (-5) + (-10)$

Use mental math to solve the equation.

26. $n + (-8) = 5$ **27.** $4 + c = 0$ **28.** $-6 + k = -14$

29. In golf, a golfer must have a score of 0 in order to be at par. A golfer scores 2 above par on the first hole, 1 below par on the second hole, and 2 below par on the third hole. Which expression can be used to decide whether the golfer is at par after the first three holes?

$(-2) + 1 + 2$ $2 + (-1) + 2$ $2 + (-1) + (-2)$

30. Copy and complete the magic square so that each row and column has a magic sum of 0. Use each integer from -4 to 4 exactly once.

3		−2
		2

Copyright © Big Ideas Learning, LLC
All rights reserved.

Name ______________________________ Date __________

11.2 Practice B

Add.

1. $15 + 24$
2. $-13 + (-35)$
3. $29 + (-29)$
4. $31 + (-72)$

5. The elevation of your plot of land is 2 feet below sea level. You add 7 feet of dirt to your land. What is the new elevation of your land?

Tell how the Commutative and Associative Properties of Addition can help you find the sum mentally. Then find the sum.

6. $18 + (-25) + (-18)$
7. $-22 + 45 + (-8)$
8. $28 + (-12) + 4$

Add.

9. $17 + (-33) + (-12)$
10. $(-41) + 25 + 19$
11. $(-43) + (-27) + 43$
12. $71 + 27 + (-42)$
13. $(-63) + 81 + 0$
14. $(-39) + (-21) + (-19)$
15. $52 + (-38) + 23$
16. $101 + (-51) + (-36)$
17. $(-117) + 125 + (-67)$

Use mental math to solve the equation.

18. $n + (-20) = 5$
19. $c + (-71) = 0$
20. $-30 + k = -110$

21. Write three integers that do not all have the same sign that have a sum of -20. Write three integers that do not all have the same sign that have a sum of 10.

22. The temperature at 6 A.M. is $-12°F$. During the next twelve hours, the temperature increases $25°F$. During the following 5 hours, the temperature decreases $23°F$. What is the temperature at 11 P.M.?

23. Copy and complete the magic square so that each row and column has a magic sum of 0. Use integers from -9 to 9, without repeating an integer.

9		−3
		2

Copyright © Big Ideas Learning, LLC
All rights reserved.

Name____ Date__________

11.2 Enrichment and Extension

Magic Squares with Integers

According to a legend, the Chinese Emperor Yu-Huang saw a magic square on the back of a turtle. In a *magic square*, the sum of the numbers in each row, column, and diagonal are the same. This sum is called the magic sum.

This magic square uses integers -6 to 2 exactly once. The magic sum is -6.

1	−6	−1
−4	−2	0
−3	2	−5

Diagonal 1: $-3 + (-2) + (-1) = -6$

Row 1: $1 + (-6) + (-1) = -6$

Row 2: $-4 + (-2) + 0 = -6$

Row 3: $-3 + 2 + (-5) = -6$

Diagonal 2: $1 + (-2) + (-5) = -6$

Column 1: $1 + (-4) + (-3) = -6$

Column 2: $-6 + (-2) + 2 = -6$

Column 3: $-1 + 0 + (-5) = -6$

Complete the magic square using each integer only once. The magic sum is given.

1. Use -9 to -1; Magic Sum $= -15$

−8		−4
	−7	

2. Use -5 to 3; Magic Sum $= -3$

−2		
		1
	−5	

3. Use -7 to 8; Magic Sum $= 2$

	7		−4
		−1	
0		3	−3
	−5		

4. Use -10 to 5; Magic Sum $= -10$

−4	1		
−9		−3	0
	−8		−6
		4	

5. Create your own magic square with integers having the magic sum 6.

Copyright © Big Ideas Learning, LLC
All rights reserved.

Name ______________________ Date ________

11.2 Puzzle Time

Why Did The Golfer Wear Two Pairs Of Pants?

Write the letter of each answer in the box containing the exercise number.

Add.

1. $12 + 5$

2. $7 + (-7)$

3. $-10 + 2$

4. $9 + (-6)$

5. $-15 + 27$

6. $23 + (-23)$

7. $-17 + 12$

8. $13 + (-15)$

9. $-9 + (-9)$

10. $-14 + (-11)$

11. $12 + (-10) + 16$

12. $15 + (-15) + 12$

13. $-22 + 30 + (-26)$

14. $-8 + (-8) + (-9)$

15. $37 + (-21) + (-16)$

16. $-42 + 8 + 17$

17. $-30 + 34 + (-9)$

18. $14 + (-21) + 7$

19. $-25 + 17 + 6$

20. $-4 + (-8) + (-6)$

21. A roller coaster climbs 84 feet on the first hill then drops 60 feet down. On the second hill the roller coaster climbs another 32 feet then drops 44 feet. What is the height at the end of the second hill?

Answers

S. 18	**N.** −18
O. 12	**C.** −8
L. −17	**H.** −25
I. −5	**E.** 0
T. 17	**G.** 3
A. −2	

7	13		3	8	11	15		14	18		4	21	1		19		10	12	16	2
17	9		5	20	6															

Copyright © Big Ideas Learning, LLC
All rights reserved.

Activity 11.3 Start Thinking!

For use before Activity 11.3

You plan a hiking trip. Discuss with a partner the math involved when hiking from one point to the next.

Activity 11.3 Warm Up

For use before Activity 11.3

Subtract.

1. $45 - 11$

2. $87 - 23$

3. $91 - 14$

4. $76 - 69$

5. $87 - 29 - 13$

6. $65 - 52 - 11$

Copyright © Big Ideas Learning, LLC
All rights reserved.

Lesson 11.3 Start Thinking!

For use before Lesson 11.3

How can you tell if the difference of two integers is positive?

How can you tell if the difference of two integers is negative?

How can you tell if the difference of two integers is zero?

Lesson 11.3 Warm Up

For use before Lesson 11.3

Subtract.

1. $5 - 8$

2. $12 - (-3)$

3. $-5 - (-6)$

4. $-4 - 6$

5. $9 - 16$

6. $-7 - 10$

Copyright © Big Ideas Learning, LLC
All rights reserved.

Name____ Date__________

11.3 Practice A

Subtract.

1. $3 - 8$
2. $4 - (-5)$
3. $-6 - 4$
4. $-9 - (-6)$
5. $10 - (-9)$
6. $12 - 4$
7. $-15 - 7$
8. $-6 - (-14)$
9. $-1 - (-3)$
10. $15 - (-7)$
11. $20 - (-10)$
12. $-31 - 14$
13. You are scuba diving at -8 feet. You dive 5 feet deeper. What is your position in the water?
14. Write $7 - 3$ using addition.
15. Write $5 + (-3)$ using subtraction.

Evaluate the expression.

16. $8 - 12 - (-6)$
17. $8 - (-8) - 3$
18. $0 - (-4) - 8$
19. $9 - (-4) + 1$
20. $7 - 12 - (-4)$
21. $-11 - (-8) - (-3)$
22. $-14 - 6 - (-2)$
23. $8 + 0 - (-11)$
24. $8 + 13 - (-5)$

Use mental math to solve the equation.

25. $a - 7 = 3$
26. $b - (-8) = -3$
27. $6 - c = 10$
28. Write two different pairs of negative integers, x and y, that make the statement $x - y = 2$ true.
29. The table shows the highest and lowest elevations for two cities.

City	Highest elevation (feet)	Lowest elevation (feet)
Long Beach, CA	360	–7
New Orleans, LA	25	–8

   a. Find the range of elevations for Long Beach.

   b. Find the range of elevations for New Orleans.

   c. One of the cities has an average elevation of about 2 feet below sea level. Which city is it?

Copyright © Big Ideas Learning, LLC
All rights reserved.

Name ______________________________ Date __________

11.3 Practice B

Subtract.

1. $8 - 13$ **2.** $18 - (-11)$ **3.** $-14 - 35$ **4.** $-51 - (-36)$

5. $100 - (-91)$ **6.** $-82 - 64$ **7.** $35 - 47$ **8.** $-36 - (-54)$

9. A dolphin is at -28 feet. It swims up and jumps out of the water to a height of 8 feet. Write a subtraction expression for the vertical distance the dolphin travels.

Evaluate the expression.

10. $15 - 42 - (-36)$ **11.** $17 - (-22) - 22$ **12.** $0 - (-41) - 28$

13. $87 - (-34) + 13$ **14.** $-35 - 27 - (-14)$ **15.** $-51 - (-23) + (-16)$

16. $-14 - 63 - (-52)$ **17.** $-28 + 10 - (-121)$ **18.** $8 - (-103) - (-95)$

Use mental math to solve the equation.

19. $a - 24 = 47$ **20.** $26 - b = 9$ **21.** $c - (-15) = 38$

22. $-5 - k = -12$ **23.** $t - (-14) = -24$ **24.** $-25 - m = 28$

25. The table shows the record monthly high and low temperatures in International Falls, Minnesota.

	Jan	Feb	Mar	Apr	May	Jun	Jul	Aug	Sep	Oct	Nov	Dec
High (°F)	48	58	76	93	95	99	98	95	95	88	73	57
Low (°F)	−46	−45	−38	−14	11	23	34	30	20	2	−32	−41

a. Find the range of temperatures for each month.

b. What are the all-time high and all-time low temperatures?

c. What is the range of the temperatures in part (b)?

26. For what values of a and b is the statement $|a - b| = |a| - |b|$ false?

Copyright © Big Ideas Learning, LLC
All rights reserved.

Name____ Date__________

11.3 Enrichment and Extension

Dart Subtraction Strategy

To play the Dart Subtraction Game, you throw two darts and find your score by subtracting the value of the second dart from the value of the first dart. It is possible to land on the same value twice.

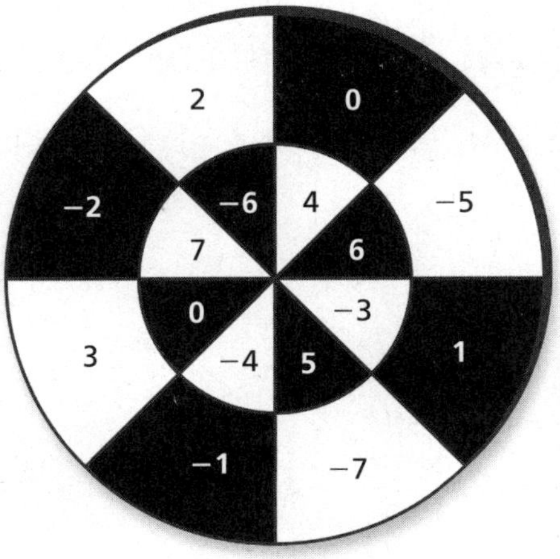

1. What is the highest possible score? What would have to be the value of the first dart? the second dart?

2. What is the lowest possible score? What would have to be the value of the first dart? the second dart?

3. List all the ways to get a score of 4. There are 11 different possibilities.

First Dart											
Second Dart											

4. List all the ways to get a score of −6.

5. How many ways could you get a score of 0? a score of 14? a score of −14?

6. If you are trying to get the *lowest* score, should you try to land on a positive or negative integer with your first dart? with your second dart? Explain your reasoning.

7. Play the game with a partner for ten rounds. Use paper punches as your darts. Hold your hand about 12 inches from the paper and drop the paper punches onto the target. Record your score after each round. Add up the score from each round to get your total score. Did you use any particular strategy? Describe.

Copyright © Big Ideas Learning, LLC
All rights reserved.

Name ______________________________ Date __________

11.3 Puzzle Time

What Did The Sea Say To The Sand?

Write the letter of each answer in the box containing the exercise number.

Subtract.

1. $3 - 11$

2. $-5 - 12$

3. $14 - (-10)$

4. $-9 - (-7)$

5. $25 - (-8)$

6. $-13 - (-13)$

Evaluate the expression.

7. $-6 + 15 - (-4)$

8. $11 - 22 - (-8)$

9. $-14 - 7 - (-25)$

10. $17 + 8 - (-15)$

11. $-9 - (-4) - 2$

12. $-16 + 5 - 12$

13. The high temperature for a day in January was 7 degrees Fahrenheit. The low temperature that day was –5 degrees Fahrenheit. What is the difference in temperatures?

14. The top of a sailboat mast is 22 feet above the water surface. The bottom of the sailboat is 3 feet below the water surface. What is the difference in the elevations?

Answers

J. 24	**H.** 12
W. −8	**G.** 40
O. 33	**E.** −7
D. −2	**I.** 0
S. 25	**N.** −17
V. −3	**A.** −23
U. 13	**T.** 4

2	5	9	13	6	2	10		6	9		3	7	14	9		1	12	8	11	4

Copyright © Big Ideas Learning, LLC
All rights reserved.

Activity 11.4 Start Thinking!

For use before Activity 11.4

How are deposits and withdrawals at a bank similar to adding integers?

Activity 11.4 Warm Up

For use before Activity 11.4

Add.

1. $9 + 9$

2. $-7 + (-7)$

3. $-3 + (-3) + (-3)$

4. $5 + 5 + 5$

5. $6 + 6 + 6 + 6$

6. $-4 + (-4) + (-4) + (-4)$

Copyright © Big Ideas Learning, LLC
All rights reserved.

Lesson 11.4

Start Thinking!

For use before Lesson 11.4

Without using your notes from Activity 11.4, explain to a partner how repeated addition is like multiplication.

Lesson 11.4

Warm Up

For use before Lesson 11.4

Multiply.

1. $7 \bullet 2$

2. $9(-7)$

3. $-6(8)$

4. $-8(-10)$

5. $6 \bullet (-5)$

6. $-5 \bullet (-12)$

Copyright © Big Ideas Learning, LLC
All rights reserved.

Name______________________________ Date__________

11.4 Practice A

Multiply.

1. $4 \bullet (-3)$
2. $-6 \bullet 5$
3. $-8(-2)$
4. $9 \bullet 6$
5. $0 \bullet (-7)$
6. $-12(-3)$
7. $11 \bullet 7$
8. $5(-5)$
9. $-13 \bullet 7$
10. $-1 \bullet 9$
11. $2(-12)$
12. $-9 \bullet (-9)$
13. A water tank leaks 5 gallons of water each day. What integer represents the change in the number of gallons of water in the tank after 7 days?

Multiply.

14. $2 \bullet (-3) \bullet 5$
15. $-5(-4)(-1)$
16. $7 \bullet 2 \bullet (-3)$
17. $0 \bullet (-8) \bullet 6$
18. $-6 \bullet 4 \bullet (-2)$
19. $5(-4)(-5)$

Evaluate the expression.

20. $(-3)^2$
21. -3^2
22. $(-2)^3$
23. -5^2
24. $-3 \bullet (-4)^2$
25. $(-7)^2 \bullet 2$
26. $|-3| \bullet (-6)$
27. $-5(-2) - 3(-4)$
28. $2 \bullet (-3)^2 - 5^2$

Find the next two numbers in the pattern.

29. 6, −12, 24, −48, …
30. 9, −27, 81, −243, …
31. An elevator is 180 feet above the first floor. Each second it descends 12 feet.

   a. What integer is the change in the height of the elevator each second?

   b. Copy and complete the table.

Time	3 sec	6 sec	9 sec
Height			

   c. Estimate how many seconds it takes the elevator to get to the first floor. Explain your reasoning.

   d. From the first floor, it takes 4 seconds to reach the basement floor. What is the height of the basement floor with respect to the first floor?

Copyright © Big Ideas Learning, LLC
All rights reserved.

Name ______________________________ Date __________

11.4 Practice B

Multiply.

1. $(-8)(-12)$
2. $10 \bullet (-14)$
3. $-21 \bullet 4$
4. $-15 \bullet (-8)$

5. The water in a pool evaporates at a rate of 16 gallons per week. What integer represents the change in the number of gallons of water in the pool after 24 weeks?

Multiply.

6. $5 \bullet (-11) \bullet (-4)$
7. $-15(-3)(-6)$
8. $-9 \bullet 0 \bullet (-3)$
9. $13 \bullet 2 \bullet (-6)$
10. $-16 \bullet 2 \bullet (-3)$
11. $-9(-9)(-9)$

Evaluate the expression.

12. $(-12)^2$
13. -12^2
14. $(-7)^3$
15. $-(-2)^3$
16. $(-2)^3 \bullet (-3)^2$
17. $(-11)^2 \bullet 7$
18. $-|-3| \bullet (-6)$
19. $11(-3) - (-2)(7)$
20. $-5 \bullet 8 - (-4)^3$

21. The gym offers a discount when more than one member of the family joins. The first member $(n = 0)$ pays \$550 per year. The second member to join $(n = 1)$ gets a discount of \$75 per year. The third member $(n = 2)$ gets an additional \$75 discount. The price for the nth member is given by $550 + (-75n)$.

    a. What is the price for the fourth member to join $(n = 3)$?

    b. For a large family, is it possible that a member would join for free? If so, which member would it be? Explain your reasoning.

    c. Other than \$0, what is the lowest amount that a member would pay to join? Which member would it be? Explain your reasoning.

22. Two integers, a and b, have a product of -48.

    a. What is the greatest possible sum of a and b?

    b. Is it possible for a and b to have a sum of 13? If so, what are the integers?

    c. What is the least possible difference of a and b?

Copyright © Big Ideas Learning, LLC
All rights reserved.

Name____ Date __________

11.4 Enrichment and Extension

Multiplying Negative Integers

In Exercises 1–4, find the product.

1. $-3 \bullet (-7) \bullet (-15) \bullet (-5)$

2. $-6 \bullet (-8) \bullet (-1) \bullet (-9)$

3. $-7 \bullet (-18) \bullet (-5) \bullet (-6) \bullet (-3)$

4. $-5 \bullet (-5) \bullet (-4) \bullet (-2) \bullet (-10)$

In Exercises 5–8, use the table.

$(-3)^2$	=	$(-3)(-3)$	=	$(-3)^1(-3)$	=	9
$(-3)^3$	=	$(-3)(-3)(-3)$	=	$(-3)^2(-3)$	=	-27
$(-3)^4$	=				=	
$(-3)^5$	=				=	
$(-3)^6$	=				=	
$(-3)^7$	=				=	
$(-3)^8$	=				=	

5. Complete the table.

6. Describe the pattern in the products.

7. Will $(-3)^{18}$ be a positive or negative integer? Explain.

8. Will $(-3)^{45}$ be a positive or negative integer? Explain.

Copyright © Big Ideas Learning, LLC
All rights reserved.

Name ______________________________ Date __________

11.4 Puzzle Time

When Do Kangaroos Celebrate Their Birthdays?

A	B	C	D	E	F		G	H	I	J		K	L	M	N

Complete each exercise. Find the answer in the answer column. Write the letter under the answer in the box containing the exercise letter.

36	G
12	O
−105	E
25	P
50	C
−60	D
72	S
45	N
−25	T
−42	U
49	P
−36	H
52	W
110	R

Multiply.

A. $5 \bullet (-12)$

B. $-14 \bullet 3$

C. $-10(-11)$

D. $8 \bullet (-7)$

E. $-9 \bullet (-5)$

F. $6(-2)(-3)$

G. $-4 \bullet 5 \bullet (-4)$

H. $(-7)(-3)(-5)$

I. $-15 \bullet 0 \bullet (-12)$

J. $(-5)^2$

K. -7^2

L. $-3^2 \bullet 8$

M. $(-4)^3$

N. You are making a necklace that is 9 inches long. You use 6 beads for each inch. What integer is the change in your supply of beads after making the necklace?

0	A
−25	M
−49	Y
−64	A
80	L
−50	U
100	B
64	F
−56	I
−110	J
66	S
−54	R
−72	E
54	K

Copyright © Big Ideas Learning, LLC
All rights reserved.

Activity 11.5

Start Thinking!

For use before Activity 11.5

What are the different ways you can multiply two integers?

Activity 11.5

Warm Up

For use before Activity 11.5

Multiply.

1. $-10 \bullet 5$

2. $-5 \bullet (-6)$

3. $8 \bullet (-9)$

4. $-15 \bullet 7$

5. $-22 \bullet (-8)$

6. $32 \bullet (-4)$

Copyright © Big Ideas Learning, LLC
All rights reserved.

Lesson 11.5 Start Thinking!
For use before Lesson 11.5

Find the next 2 numbers in the pattern: 12, 6, 24, 12, 48, ... Explain your reasoning.

Lesson 11.5 Warm Up
For use before Lesson 11.5

Divide.

1. $12 \div (-3)$

2. $-32 \div 4$

3. $-56 \div (-8)$

4. $\frac{-45}{9}$

5. $\frac{81}{-9}$

6. $\frac{-144}{-12}$

Copyright © Big Ideas Learning, LLC
All rights reserved.

Name ____________________ Date __________

11.5 Practice A

Divide, if possible.

1. $8 \div (-4)$
2. $-15 \div (-3)$
3. $\frac{-10}{5}$
4. $0 \div (-7)$
5. $-35 \div 7$
6. $\frac{18}{-6}$
7. $-72 \div 9$
8. $-5 \div 5$
9. $\frac{15}{0}$
10. $12 \div (-2)$
11. $\frac{-56}{-8}$
12. $21 \div (-3)$
13. Your team dives for 28 lobsters over 7 days. What is the average daily lobster catch?

Find the mean of the integers.

14. $5, -7, 12, -10, 15$
15. $-16, -27, 21, -19, 14, -3$

Evaluate the expression.

16. $6 - 12 \div (-3)$
17. $|-16| \div (-2)^2 - 4^2$
18. $\frac{-10 + (-2)^3}{-3}$

Find the next two numbers in the pattern.

19. $-96, 48, -24, 12, \ldots$
20. $12{,}500, -2500, 500, -100, \ldots$
21. A skateboarder descends on a ramp from 172 feet to 67 feet in 15 seconds. What is the average change in height per second?
22. The velocity (in feet per second) of a bouncing ball was recorded every second. The table shows the velocity for each second.

Time (sec)	1	2	3	4	5
Velocity (ft/sec)	–15	–6	2	10	–11

   a. What is the average velocity of the bouncing ball over the 5 seconds?

   b. What is the highest recorded speed of the bouncing ball? Is the ball going up or down at this speed?

   c. During the 5 second period, did the ball spend more time going up or going down? Explain your reasoning.

   d. Between which two seconds did the ball change from going up to going down? Explain your reasoning.

Copyright © Big Ideas Learning, LLC
All rights reserved.

Name ______________________________ Date __________

11.5 Practice B

Divide, if possible.

1. $51 \div (-3)$

2. $\frac{-63}{21}$

3. $\frac{0}{25}$

4. $\frac{-144}{-9}$

5. $105 \div (-5)$

6. $-82 \div 0$

7. $-96 \div 8$

8. $-15 \div (-15)$

9. $\frac{99}{-9}$

10. $225 \div (-25)$

11. $\frac{-156}{3}$

12. $-48 \div (-3)$

13. Your team catches 42 Mahi Mahi over 2 weeks. What is the average daily Mahi Mahi catch?

Evaluate the expression.

14. $-10 + 16 \div (-2) + 7$

15. $(-68) \div (-4) + 5 \bullet (-3)$

16. $10 - 12^2 \div (-2)^3$

17. $-|-16| \div (-8) \bullet 5^2$

18. $\frac{3 + 7 \bullet (-3^2)}{-5}$

19. PI-Squared and Euler Circles are in a math competition consisting of 10 two-part questions. Both parts correct earns 5 points, one part correct earns 2 points, and no parts correct earns −1 point.

Team	Both	One	None
PI-Squared	4	2	4
Euler Circles	2	6	2

a. What is the mean points per question for PI-Squared?

b. What is the mean points per question for Euler Circles?

c. Which team should win the competition? Explain your reasoning.

20. A 155-pound person burns about 500 calories per hour playing racquetball.

a. One pound is equal to 3500 calories. How long will it take to burn 1 pound playing racquetball?

b. How long will it take to burn 5 pounds playing racquetball? Explain your reasoning.

c. If the person were to rest 5 minutes for every 30 minutes of playing, how long would it take to burn 1 pound?

Copyright © Big Ideas Learning, LLC
All rights reserved.

Name____________________ Date__________

11.5 Enrichment and Extension

Absolute Value and Integers

Assume $a > 0$ and $b < 0$. Determine whether the statement is *always, sometimes,* or *never* true.

1. $a + b > 0$ **2.** $a + b < 0$ **3** $|a| + |b| < 0$ **4.** $|a| + |b| > 0$

5. $a - b > 0$ **6.** $a - b < 0$ **7.** $b - a > 0$ **8.** $b - a < 0$

9. $|a| - |b| > 0$ **10.** $|a| - |b| < 0$ **11.** $-|a| - |b| < 0$ **12.** $|a| - (-|b|) < 0$

13. $ab > 0$ **14.** $ab < 0$ **15.** $\frac{a}{b} < 0$ **16.** $\frac{a}{b} > 0$

Assume a, b, and n are integers, with $a > 0$ and $b < 0$. Determine whether the possible values of n are *all integers, all positive integers and zero, all negative integers and zero,* or *none.*

17. $-|-n| - |n| > 0$ **18.** $-|n| + |-n| = 0$ **19.** $|n| + (-n) = 0$ **20.** $-n + n = 0$

21. $an \geq 0$ **22.** $an \leq 0$ **23.** $\frac{n}{b} \geq 0$ **24.** $\frac{n}{b} \leq 0$

25. $|a| + |n| = a + n$ **26.** $|a| + |n| = a - n$ **27.** $|an| = |a| \bullet |n|$ **28.** $\left|\frac{n}{b}\right| = \frac{|n|}{|b|}$

29. $\left|\frac{n}{a}\right| = \frac{n}{a}$ **30.** $|an| = an$ **31.** $|bn| = bn$ **32.** $\left|\frac{n}{b}\right| = \frac{n}{b}$

For each exercise number, use your answers and the key below to color the cell. Do not color the cells that have a zero in them.

Always = Purple

Sometimes = Blue

Never = Red

All Integers = Blue

All Positive Integers and Zero = Orange

All Negative Integers and Zero = Red

None = Purple

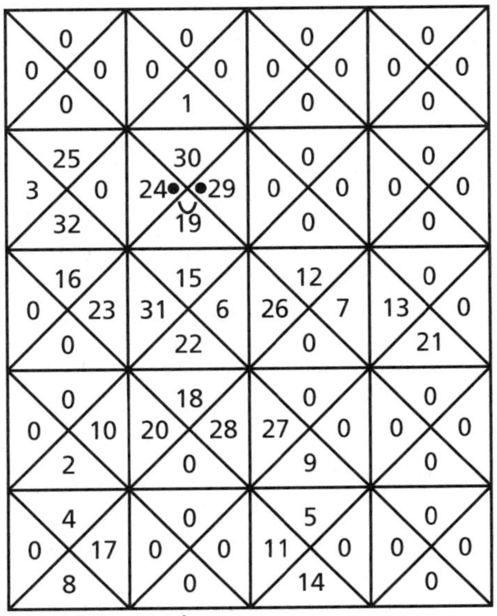

©Buki Toys Ltd.

Copyright © Big Ideas Learning, LLC
All rights reserved.

Name ______________________________________ Date __________

11.5 Puzzle Time

What Did The Baseball Mitt Say To The Ball?

Circle the letter of each correct answer in the boxes below. The circled letters will spell out the answer to the riddle.

Divide.

1. $6 \div (-3)$

2. $-52 \div (-4)$

3. $-27 \div 3$

4. $-36 \div 2$

5. $56 \div (-8)$

6. $-24 \div (-3)$

7. $\dfrac{-18}{6}$

8. $\dfrac{25}{-5}$

9. $\dfrac{-16}{-4}$

10. $\dfrac{-66}{11}$

Evaluate the expression.

11. $32 \div (-2) + (-25) \div 5$

12. $4 \bullet (-3) + 12 \div (-4)$

13. You improve your time running a course by 5 seconds in week one, by 3 seconds in week two, and by 4 seconds in week three. What is the average weekly change in your running time?

B	C	A	R	T	S	C	H	E	Y	D	O	U	N	L	A	S	T	E	O	R
14	–21	4	12	13	2	8	–9	20	–7	15	–15	–3	9	–6	–18	–8	–2	–5	–13	–4

Copyright © Big Ideas Learning, LLC
All rights reserved.

Name______________________________ Date__________

Chapter 11

Technology Connection

For use after Section 11.2

Entering Negative Numbers on a Calculator

You can use the [+ | −] key on a calculator to enter negative numbers.

EXAMPLE 1 How does your calculator display negative numbers? Find the value of 0 − 3.

SOLUTION

Press 0 [−] 3 [=].

The integer that is 3 less than 0 is −3. Study the calculator display and determine where the negative sign is displayed. It may be immediately to the left of the 3, or it may be at the far left of the display.

Displays

[−3] or [− 3]

Now clear the display and try entering −2 using the [+ | −] key, also called the change-sign key. Press 2 [+ | −]. The display should be similar to the one you found in the first example. A negative sign displayed to the far left can be easy to miss if you are not looking for it; this will make a big difference in your calculations.

EXAMPLE 2 The temperature is −2°. It rises 6°. What is the new temperature?

SOLUTION

You need to find −2 + 6.

Press 2 [+ | −] + 6 [=]. The answer is 4.

Use the [+ | −] key on a calculator to evaluate the expression.

1. 0 − 8

2. −5 + 5

3. −1 + 0

4. 1 − 2

5. 0 − 7

6. −4 + 2

7. What do you think happens when you subtract 1 from a negative number? Explain your reasoning. Check your answer with a calculator.

Copyright © Big Ideas Learning, LLC
All rights reserved.

Chapter 12

Copyright © Big Ideas Learning, LLC
All rights reserved.

Name______________________________ Date __________

Chapter 12 Rational Numbers

Dear Family,

Every family has a favorite pizza. Pizzas are unusual in that the ingredients are normally customized. The variety of meats, vegetables, sauces, crusts, herbs and spices produces a dizzying number of possibilities.

You and your student might enjoy making homemade pizza. First, you have to decide how many pizzas you need and how big each will be. If there are just two people eating, a small pizza might be just right. But for a large group, several pizzas may be needed.

You might estimate that each person will eat about one fourth of a large pizza. Multiply that by the number of people eating and you know how many fourths are needed. From that, work with your student to figure out how many whole pizzas are needed.

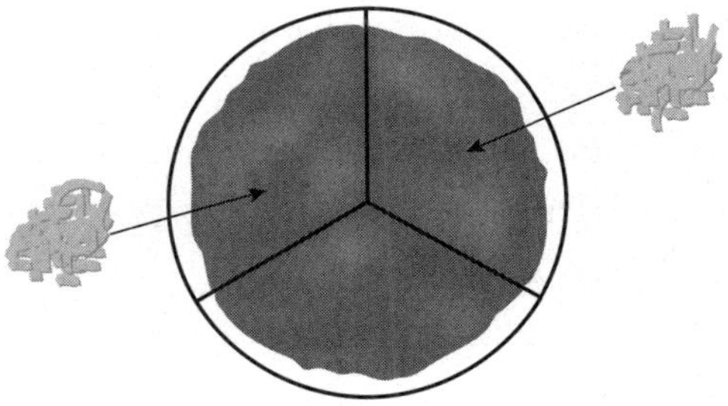

Next you have to choose and prepare the toppings—maybe even customize the toppings for each person. As you choose the toppings, divide them so they will be evenly distributed. If two thirds of your group wants green peppers on the pizza, divide the diced green peppers into two equal piles. Divide the pizza into three equal parts and put the piles on two of the parts. Talk with your student about a strategy for doing this when you have more than one pizza.

Finally, decide how you will cut the pizza. If there are two people, you probably will choose an even number of slices. If there are five people, you might choose to cut the pizza into ten slices, so that each person can have two slices. Count the number in your group and think of a good number of slices to use. What if you have more than one pizza?

Enjoy your pizza—and don't hold the math!

Copyright © Big Ideas Learning, LLC
All rights reserved.

Nombre ______________________________ Fecha ________

Capítulo 12 Números Racionales

Estimada Familia:

Cada familia tiene una pizza favorita. Las pizzas no son comunes en el sentido en que los ingredientes generalmente son personalizados. La variedad de carnes, verduras, salsas, masas, hierbas y especies produce un abrumador número de posibilidades.

Usted y su estudiante pueden disfrutar preparando pizza en casa. Primero, tienen que decidir cuántas pizzas necesitan y qué tan grande será cada una. Si sólo dos personas van a comer una pizza pequeña será suficiente. Pero para un grupo grande, se necesitarán varias pizzas.

Pueden estimar que cada persona comerá alrededor de un cuarto de una pizza grande. Multipliquen eso por el número de personas que van a comer y entonces sabrán cuántos cuartos necesitarán. A partir de ese dato, trabaje con su estudiante para averiguar cuántas pizzas enteras se necesitarán.

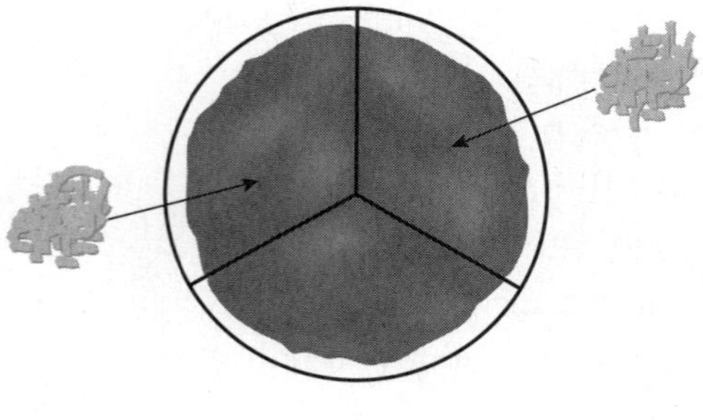

Luego, tendrán que elegir y preparar las cubiertas—quizás incluso personalizar las cubiertas para cada persona. A medida que eligen las cubiertas, divídanlas para que queden distribuidas de manera uniforme. Si dos tercios de su grupo desean pimientos verdes en la pizza, dividan los pimientos verdes picados en dos montones iguales. Dividan la pizza en tres partes iguales y coloquen los montones en dos de las partes. Converse con su estudiante acerca de una estrategia para hacer esto cuando tienen más de una pizza.

Finalmente, decidan cómo cortarán la pizza. Si hay dos personas, probablemente elegirán un igual número de tajadas. Si son cinco personas, querrán cortar la pizza en diez tajadas, para que cada persona reciba dos. Cuenten el número de personas en su grupo y piensen acerca de un buen número de tajadas para usar. ¿Qué pasa si tienen más de una pizza?

¡Disfruten su pizza—y no dejen de usar las matemáticas!

Copyright © Big Ideas Learning, LLC
All rights reserved.

Activity 12.1 Start Thinking!
For use before Activity 12.1

You eat $\frac{3}{8}$ of a pie and your sister eats $\frac{1}{4}$ of the pie. Explain who ate more of the pie.

Activity 12.1 Warm Up
For use before Activity 12.1

Copy and complete the statement using <, >, or =.

1. $\frac{2}{3}$ $\underline{\quad ? \quad}$ $\frac{4}{6}$

2. 0.7 $\underline{\quad ? \quad}$ $\frac{3}{4}$

3. 1.5 $\underline{\quad ? \quad}$ $\frac{2}{3}$

4. -0.6 $\underline{\quad ? \quad}$ -1

5. $3\frac{7}{10}$ $\underline{\quad ? \quad}$ $3\frac{4}{5}$

6. -2.4 $\underline{\quad ? \quad}$ $-2\frac{2}{5}$

Copyright © Big Ideas Learning, LLC
All rights reserved.

Lesson 12.1 Start Thinking!

For use before Lesson 12.1

You and two friends are playing basketball. You make 7 out of 15 shots. Your first friend makes 6 out of 10 shots and your second friend makes 5 out of 12 shots. Who is the better shooter?

How would you solve this problem using what you know about rational numbers?

Lesson 12.1 Warm Up

For use before Lesson 12.1

Order the numbers from least to greatest.

1. $-\frac{2}{3}, 0.6, \frac{3}{4}, -\frac{7}{4}, -0.3$

2. $1.5, -1.3, \frac{7}{5}, -\frac{6}{5}, 1.65$

3. $-2.75, \frac{11}{4}, \frac{5}{4}, -0.37, 2.65$

4. $\frac{4}{10}, -0.8, \frac{1}{8}, 4.5, -\frac{4}{2}$

5. $3.8, -\frac{9}{3}, -0.3, \frac{6}{4}, \frac{8}{5}$

6. $-1.5, \frac{7}{3}, -\frac{3}{4}, 0.6, \frac{9}{6}$

Copyright © Big Ideas Learning, LLC
All rights reserved.

Name______________________________ Date__________

12.1 Practice A

Write the rational number as a decimal.

1. $\frac{5}{9}$
2. $-\frac{3}{8}$
3. $-\frac{3}{11}$
4. $\frac{7}{30}$
5. $1\frac{5}{12}$
6. $-2\frac{1}{3}$
7. $-\frac{13}{22}$
8. $5\frac{1}{6}$

Write the decimal as a fraction or mixed number in simplest form.

9. 0.7
10. –0.3
11. –0.43
12. 0.52
13. 1.25
14. –2.07
15. 4.18
16. 3.125

Order the numbers from least to greatest.

17. $1.6, -\frac{2}{3}, -0.5, \frac{3}{2}, -\frac{5}{2}$
18. $\frac{3}{4}, -1.7, 0.6, -\frac{7}{4}, 1.1$
19. $0, -\frac{2}{5}, 0.67, \frac{7}{9}, -0.5$
20. $-\frac{1}{3}, -0.3, \frac{4}{3}, 1.2, \frac{3}{2}$

21. You receive two quarters, one dime, and four pennies back in change.

 a. Write the amount as a decimal.

 b. Write the amount as a fraction in simplest form.

22. In football, a completion percentage is the number of completions divided by the number of passes. Does Tom or Ian have a higher completion percentage?

Player	Passes	Completions
Tom	22	10
Ian	18	9

23. You get 17 out of 22 questions correct on a math test.

 a. What is your percent of correct answers?

 b. The lowest score to pass is 70%. Did you pass the test?

 c. What is the minimum number of correct answers needed in order to pass the test?

Copyright © Big Ideas Learning, LLC
All rights reserved.

Name ______________________________ Date __________

12.1 Practice B

Write the rational number as a decimal.

1. $\frac{5}{8}$ **2.** $-\frac{3}{22}$ **3.** $1\frac{2}{9}$ **4.** $-5\frac{3}{40}$

5. $-7\frac{5}{11}$ **6.** $4\frac{1}{15}$ **7.** $-9\frac{1}{9}$ **8.** $-7\frac{5}{6}$

Write the decimal as a fraction or mixed number in simplest form.

9. 0.68 **10.** −0.01 **11.** −3.99 **12.** 8.745

13. 3.005 **14.** −13.012 **15.** −9.98 **16.** −10.452

17. You caught a red snapper that is $8\frac{5}{12}$ inches long. Your friend caught a red snapper that is $8\frac{6}{13}$ inches long. Who caught the larger red snapper?

Copy and complete the statement using <, >, or =.

18. $0.13 \underline{\ ?\ } \frac{1}{8}$ **19.** $-1\frac{2}{9} \underline{\ ?\ } -\frac{5}{4}$ **20.** $-5.175 \underline{\ ?\ } -5\frac{1}{6}$

21. Find one terminating decimal and one repeating decimal between $-1\frac{1}{2}$ and $-1\frac{7}{9}$.

22. The table gives the tidal changes in the water level of a lagoon for every six hours of a given day.

Time	4:00 A.M.	10:00 A.M.	4:00 P.M.	10:00 P.M.
Change (feet)	2.25	$-2\frac{6}{7}$	$-\frac{3}{2}$	$2\frac{1}{3}$

a. Order the numbers from least to greatest.

b. At what time(s) did the water level decrease?

c. What was the largest change in water level?

d. Did the tidal change in part (c) involve an increase or a decrease in water level?

e. Will the next tidal change be an increase or decrease in water level? Explain.

Copyright © Big Ideas Learning, LLC
All rights reserved.

Name _______________________________________ Date __________

12.1 Enrichment and Extension

Where do I Belong?

An **irrational number** is a decimal that goes on forever and does not repeat. Irrational numbers cannot be written as fractions.

Examples: 3.14159..., −0.010110111..., 2.2360679...

A **real number** is any number that is either rational or irrational.

The Venn diagram shows how real numbers, rational numbers, irrational numbers, integers, and whole numbers are related.

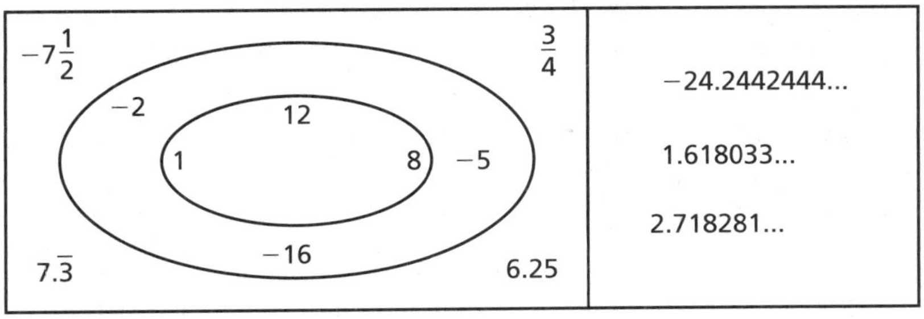

1. Describe any patterns you notice with the numbers.
2. Place the name of each number type in its appropriate spot.

   Real number   Rational number   Irrational number   Integer   Whole number

For each real number, tell whether it is *rational*, *irrational*, an *integer*, and/or a *whole number*. You may have more than one answer.

3. $4\frac{2}{3}$ 4. $-32.\overline{3}$ 5. 0.919911...   6. 15

7. Can you write 2 as a fraction? Are all whole numbers also rational numbers?
8. Can you write −14 as a fraction? Are all integers also rational numbers?

Complete the statement with *always*, *sometimes*, or *never*.

9. A **real number** chosen at random is __?__ an **integer**.
10. An **integer** chosen at random is __?__ a **real number**.
11. An **irrational number** chosen at random is __?__ a **rational number**.
12. A **rational number** chosen at random is __?__ an **integer**.
13. A **whole number** chosen at random is __?__ a **rational number**.

Copyright © Big Ideas Learning, LLC
All rights reserved.

Name ______________________________ Date __________

12.1 Puzzle Time

Did You Hear About...

A	B	C	D	E	F
G	H	I	J	K	L
M	N	O	P	Q	R

Complete each exercise. Find the answer in the answer column. Write the word under the answer in the box containing the exercise letter.

$4.1\overline{6}$	WRITE
-0.375	STUDENT
-3.875	COULDN'T
$-0.41\overline{6}$	WHO
0.125	WATERPROOF
$5\frac{11}{200}$	HAVE
$\frac{27}{40}$	GOLDFISH
$-1\frac{13}{50}$	HE
$-\frac{7}{10}$	ESSAY

Write the rational number as a decimal.

A. $\frac{8}{9}$

B. $-\frac{3}{8}$

C. $-\frac{5}{12}$

D. $\frac{23}{30}$

E. $1\frac{3}{4}$

F. $-3\frac{7}{8}$

G. $4\frac{1}{6}$

H. $4\frac{4}{25}$

Write the decimal as a fraction or mixed number in simplest form.

I. -0.7

J. 0.84

K. 0.675

L. -0.252

M. -1.26

N. -2.78

O. 5.055

P. -11.688

Q. You eat one slice of a pizza that is cut into 8 even slices. What is the amount you ate written as a decimal?

R. At basketball practice, Charlie makes 52 baskets out of 80 shots. What percentage of baskets did he make?

1.75	HE
$0.7\overline{6}$	SAID
65%	INK
$0.\overline{8}$	THE
4.16	HIS
$-2\frac{39}{50}$	DIDN'T
$-11\frac{86}{125}$	ANY
$-\frac{63}{250}$	BECAUSE
$\frac{21}{25}$	ON

Copyright © Big Ideas Learning, LLC
All rights reserved.

Activity 12.2 Start Thinking!

For use before Activity 12.2

The temperature on a given day is 55°. Explain to a partner how to use addition to find how the temperature would change if the temperature in one day increased 15 degrees, decreased 20 degrees, and then increased 8 degrees.

Is the temperature at the end of the day greater or less than the temperature at the beginning of the day? Explain.

Activity 12.2 Warm Up

For use before Activity 12.2

Add.

1. $-54 + (-23)$

2. $78 + (-24)$

3. $-23 + 65$

4. $-45 + 25$

5. $62 + (-29)$

6. $-87 + (-12)$

Copyright © Big Ideas Learning, LLC
All rights reserved.

Lesson 12.2 Start Thinking!

For use before Lesson 12.2

Explain how to find $\frac{4}{5} + \left(-\frac{2}{5}\right)$.

Explain how to find $-2.6 + 5.8$.

Lesson 12.2 Warm Up

For use before Lesson 12.2

Add. Write fractions in simplest form.

1. $3\frac{4}{9} + \left(-\frac{2}{3}\right)$

2. $-7 + 4\frac{5}{9}$

3. $-2\frac{2}{5} + 1\frac{2}{3}$

4. $-12.3 + 5.4$

5. $1.6 + (-19.8)$

6. $5.3 + (-7.8)$

Copyright © Big Ideas Learning, LLC
All rights reserved.

Name ______________________________ Date __________

12.2 Practice A

Add. Write fractions in simplest form.

1. $\frac{5}{16} + \left(-\frac{7}{16}\right)$
2. $\frac{3}{5} + \left(-\frac{4}{15}\right)$
3. $-\frac{7}{2} + 3\frac{2}{3}$
4. $5.6 + (-1.3)$
5. $-8.2 + 5.4$
6. $7.15 + (-12.76)$
7. Describe and correct the error in finding the sum.

$$\times \quad \frac{3}{10} + \left(-\frac{1}{10}\right) = \frac{3+1}{10} = \frac{4}{10} = \frac{2}{5}$$

Evaluate the expression when $x = \frac{1}{2}$ and $y = -\frac{2}{5}$.

8. $-x + y$
9. $x + 2y$
10. $|x + y|$
11. The temperature is -12.6 degrees Celsius. The temperature goes up 7.9 degrees. What is the new temperature?
12. You finish $\frac{3}{8}$ of the project. Your friend finishes $\frac{1}{4}$ of the project. What fraction of the project is finished?

Add. Write fractions in simplest form.

13. $5 + \left(-2\frac{1}{3}\right) + \left(-3\frac{1}{6}\right)$
14. $-4\frac{1}{5} + 3\frac{2}{3} + \left(-1\frac{2}{5}\right)$
15. $-12.4 + 19.1 + (-4.3)$
16. Determine if the following statements are *always*, *sometimes*, or *never* true.
    a. When adding two negative rational numbers, the sum will be negative.
    b. When adding two rational numbers with different signs, the sum will be zero.
    c. When adding two positive rational numbers, the sum will be zero.
    d. When adding two rational numbers with different signs, the sum will be negative.

Copyright © Big Ideas Learning, LLC
All rights reserved.

Name ______________________________ Date ________

12.2 Practice B

Add. Write fractions in simplest form.

1. $\frac{2}{5} + \left(-\frac{3}{15}\right)$

2. $\frac{7}{12} + \left(-1\frac{2}{3}\right)$

3. $\frac{2}{7} + \left(-3\frac{5}{14}\right)$

4. $7.26 + (-13.43)$

5. $-18.02 + 15.68$

6. $-15.75 + (-12.76)$

7. Describe and correct the error in finding the sum.

$$\times \quad 2\frac{5}{6} + \left(-\frac{8}{15}\right) = \frac{13}{6} + \left(-\frac{8}{15}\right) = \frac{65 + (-16)}{30} = \frac{49}{30} = 1\frac{19}{30}$$

Evaluate the expression when $x = -\frac{1}{5}$ and $y = \frac{3}{4}$.

8. $x + (-y)$

9. $4x + y$

10. $-|x| + y$

11. Your banking account balance is $-\$1.56$. You deposit \$10. What is your new balance?

12. You mow $\frac{1}{3}$ of the lawn. Your sister mows $\frac{2}{7}$ of the lawn. What fraction of the lawn is mowed?

Add. Write fractions in simplest form.

13. $1\frac{1}{4} + \left(-4\frac{1}{5}\right) + \left(-2\frac{3}{5}\right)$

14. $-\frac{1}{3} + 2\frac{2}{9} + \left(-5\frac{2}{3}\right)$

15. $-1.5 + (14.2) + 7.3$

16. When is the sum of two rational numbers with different signs positive?

17. The table at the right shows the amount of snowfall (in inches) for three months compared to the yearly average. Is the snowfall for the three-month period greater than or less than the yearly average? Explain.

December	January	February
$1\frac{2}{3}$	$-2\frac{1}{6}$	$2\frac{5}{6}$

18. The table below shows the weekly profits of a concession stand. What must the Week 5 profit be to break even over the 5-month period?

Week 1	Week 2	Week 3	Week 4	Week 5
2.4	−1.7	5.4	−3.75	?

Copyright © Big Ideas Learning, LLC
All rights reserved.

Name____ Date__________

12.2 Enrichment and Extension

Fun with Puzzles

3, 3, 4, 9	1, 2, 3, 5	4, 6, 7, 8	2, 3, 7, 8
3, 4, 5, 6	2, 3, 8, 9	1, 2, 6, 9	1, 2, 3, 4

Choose a set of numbers from above, and fill in the boxes to make each equation true. Each set of numbers will be used once.

1. $\square\frac{\square}{\square} - 4\frac{\square}{2} = -1\frac{1}{10}$

2. $-\square.5\square + \square7.\square = 35.32$

3. $\square\square.\square - 4\square8 = -410.2$

4. $-\frac{\square}{\square} - \square\frac{\square}{2} = -3\frac{1}{4}$

5. $-\square.8\square - \square2.\square = -46.19$

6. $\square\frac{\square}{8} - \square\frac{3}{\square} = -1\frac{7}{8}$

7. $-\square\frac{\square}{3} + 2\frac{\square}{\square} = -4\frac{5}{9}$

8. $-\square8 + \square\square.\square = 18.4$

9. Use the numbers 1 through 8 to fill in the blanks and make the equation true. There is more than one way to do it.

$\square.\square\square - \square.\square\square = -1.\square\square$

Copyright © Big Ideas Learning, LLC
All rights reserved.

Name ______________________________ Date __________

12.2 Puzzle Time

Where Do Polar Bears Vote?

Write the letter of each answer in the box containing the exercise number.

Add. Write fractions in simplest form.

1. $\frac{5}{6} + \frac{8}{6}$

2. $\frac{7}{10} + \left(-\frac{3}{5}\right)$

3. $-\frac{9}{2} + \frac{5}{12}$

4. $5\frac{1}{3} + \left(-\frac{5}{9}\right)$

5. $\frac{3}{5} + \frac{8}{5}$

6. $-4 + \frac{3}{2}$

7. $3.6 + (-2.4)$

8. $-8.2 + 9.1$

9. $6.8 + (-3.2)$

10. $-4.5 + (-4.7)$

11. $5.327 + (-2.25)$

12. $14.62 + (-11.302)$

13. Sara has $4\frac{3}{4}$ yards of red fleece and $2\frac{2}{3}$ yards of blue fleece fabric. How many yards of red and blue fleece fabric does she have altogether?

14. On Saturday, you biked 7.5 miles. On Sunday, you biked 8.9 miles. How many miles did you bike altogether?

Answers

O. $2\frac{1}{6}$

T. 3.6

E. $-2\frac{1}{2}$

O. $2\frac{1}{5}$

P. 3.077

L. 16.4

T. $\frac{1}{10}$

H. −9.2

R. $7\frac{5}{12}$

A. $4\frac{7}{9}$

E. $-4\frac{1}{12}$

T. 1.2

N. 3.318

H. 0.9

4	9		2	8	3		12	1	13	7	10		11	5	14	6

Copyright © Big Ideas Learning, LLC
All rights reserved.

Activity 12.3 Start Thinking!

For use before Activity 12.3

Think of some sports where you add and subtract rational numbers. Give examples of why you would add rational numbers. Give examples that use negative rational numbers.

Activity 12.3 Warm Up

For use before Activity 12.3

Subtract.

1. $-45 - 25$
2. $62 - (-29)$
3. $-87 - (-12)$
4. $-32 - 43$
5. $-75 - (-87)$
6. $-12 - (-54)$

Copyright © Big Ideas Learning, LLC
All rights reserved.

Lesson 12.3 Start Thinking!

For use before Lesson 12.3

Explain how to find $-\frac{4}{5} - \frac{2}{5}$.

Explain how to find $-2.6 - 5.8$.

Lesson 12.3 Warm Up

For use before Lesson 12.3

Subtract. Write fractions in simplest form.

1. $-7 - \frac{5}{7}$

2. $-5\frac{9}{10} - 7\frac{3}{5}$

3. $-4\frac{1}{4} - 6\frac{3}{8}$

4. $-13 - 5.9$

5. $14.6 - (-9.2)$

6. $-7.4 - 10.6$

Copyright © Big Ideas Learning, LLC
All rights reserved.

Name ____________________ Date __________

12.3 Practice A

Subtract. Write fractions in simplest form.

1. $\frac{3}{7} - \frac{10}{7}$

2. $\frac{7}{12} - \left(-\frac{13}{12}\right)$

3. $-\frac{1}{3} - \left(-\frac{9}{4}\right)$

4. $-3\frac{1}{2} - 1\frac{5}{6}$

5. $-12.41 - (-9.95)$

6. $2 - 8.25$

Find the distance between the two numbers on a number line.

7. $6, -4\frac{1}{4}$

8. $-3.1, -5.7$

9. $-1\frac{1}{3}, -4\frac{2}{5}$

10. Your dog's water bowl is $\frac{3}{4}$ full. After taking a drink, the water bowl is $\frac{1}{3}$ full. What fraction of the bowl did your dog drink?

Evaluate.

11. $\frac{7}{8} - \left(-2\frac{3}{4}\right) + \left(-4\frac{1}{2}\right)$

12. $5.76 - (-2.31) - 10.64$

13. Mary filled a water cooler with $6\frac{1}{2}$ gallons of water. She forgot to close the plug and $2\frac{5}{6}$ gallons leaked out.

a. How many gallons of water remain in the cooler?

b. She adds $1\frac{1}{4}$ gallons. How many gallons of water are now in the cooler?

c. How many gallons of water must she add to the cooler to get the required $6\frac{1}{2}$ gallons?

14. Is the difference of two positive rational number always positive? Explain.

Copyright © Big Ideas Learning, LLC
All rights reserved.

Name ______________________________ Date __________

12.3 Practice B

Subtract. Write fractions in simplest form.

1. $\frac{7}{3} - \frac{8}{15}$

2. $-\frac{7}{24} - \left(-\frac{5}{8}\right)$

3. $1\frac{5}{6} - \left(-2\frac{1}{4}\right)$

4. $-3\frac{7}{8} - 9\frac{5}{6}$

5. $-102.431 - (-59.95)$

6. $12.001 - 8.215$

Find the distance between the two numbers on a number line.

7. $-7\frac{1}{5}, -4\frac{2}{3}$

8. $-9.2, 4.5$

9. $-2, -3.7$

10. The largest orange in a bag has a circumference of $9\frac{5}{8}$ inches. The smallest orange has a circumference of $7\frac{13}{16}$ inches. Write the difference of the circumferences of the smallest orange and the largest orange.

Evaluate.

11. $\frac{5}{12} - \left(-3\frac{1}{4}\right) + \left(-6\frac{1}{2}\right) - 3$

12. $23.706 - (-82.31) - 130.641$

13. $-\frac{3}{8} - (-4.35)$

14. $-\frac{5}{18} - \left|-\frac{1}{6}\right| + \left(-\frac{7}{9}\right)$

15. Your bank account balance is \$32.00. You make the following withdrawals, in the following order: \$15.00, \$7.41, \$35.79, and \$0.53. After each withdrawal that leaves a negative balance, the bank adds a –\$32.00 bank fee to your account. What is your new balance?

16. Fill in the blanks to make the solution correct.

$$3\frac{3}{4} - \square\frac{\square}{8} = 2$$

Copyright © Big Ideas Learning, LLC
All rights reserved.

Name______________________________ Date__________

12.3 Enrichment and Extension

Adding and Subtracting Numbers

Use the information to label the total change.

1. In golf, each hole has a par, or the amount of strokes it should take to complete the hole. Find the total number of strokes above or below par the golfer shot after the first six holes golfing.

Par	5	3	5	5	5	4
# of Strokes	3	7	3	6	6	6

2. Find the amount of money left in the account.

Transaction	Credit	Debit	Balance
Beginning Balance			37.07
Deposit	38.10		
Withdrawal		8.79	
Withdrawal		63.12	

3. Find the total number of yards the football team gained or lost.

First down: Loss of 17 yards

Second down: Gain of 5 yards

Third down: Gain of 9 yards

Fourth down: Gain of 1 yard

Determine if the situation would result in a *positive number*, a *negative number* or *zero*. Explain.

4. Your bank account has a balance of $111.13. You write a check for $113.10.

5. You walk to your friends' house then back home. Your friend lives $\frac{7}{10}$ of a mile away.

6. The temperature at the beginning of the day was 9° Fahrenheit. The temperature dropped 14° Fahrenheit.

7. You cut a piece of wood into a $3\frac{1}{3}$-foot piece and a $2\frac{2}{3}$-foot piece. The original length of the board was 6 feet.

Copyright © Big Ideas Learning, LLC
All rights reserved.

Name ____ Date __________

12.3 Puzzle Time

Where Does A Salad Dressing Get A Good Night's Sleep?

Write the letter of each answer in the box containing the exercise number.

Subtract. Write the fractions in simplest form.

1. $\frac{3}{4} - \frac{9}{4}$

2. $-3 - \frac{7}{2}$

3. $-\frac{1}{5} - \left(-\frac{5}{11}\right)$

4. $-\frac{5}{8} - \frac{2}{7}$

5. $-2\frac{2}{3} - 4\frac{1}{6}$

6. $-3\frac{1}{9} - \left(-2\frac{1}{3}\right)$

7. $-7 - 3.2$

8. $6.1 - 5.8$

9. $-4.125 - (-2.8)$

10. $-12.33 - 7.21$

11. $5.67 - (-3.142)$

12. $2.567 - 6.814$

Find the distance between the two numbers on a number line.

13. $-3\frac{1}{4}, 4\frac{1}{2}$

14. $-6.1, 8.4$

15. Your project requires a board that has a length of $5\frac{3}{16}$ inches. You found a board that has a length of $9\frac{1}{8}$ inches. How much of the board needs to be cut to use it for your project?

Answers

O. $-\frac{7}{9}$	**A.** $-6\frac{1}{2}$
T. 8.812	**E.** $-1\frac{1}{2}$
O. −10.2	**B.** −4.247
T. 0.3	**E.** $\frac{14}{55}$
C. $7\frac{3}{4}$	**F.** −1.325
L. $3\frac{15}{16}$	**E.** $-\frac{51}{56}$
D. 14.5	**N.** −19.54
U. $-6\frac{5}{6}$	

6	10		2		12	4	14		7	9		15	3	11	8	5	13	1

Copyright © Big Ideas Learning, LLC
All rights reserved.

Activity 12.4

Start Thinking!

For use before Activity 12.4

Draw a diagram to represent the following football plays: gain of 2 yards, loss of 8 yards, gain of 20 yards, loss of 5 yards, and gain of 3 yards.

Activity 12.4

Warm Up

For use before Activity 12.4

Multiply.

1. $-12 \bullet 9$ **2.** $11(-10)$ **3.** $14 \bullet 12$

Divide.

4. $\dfrac{-48}{-6}$ **5.** $\dfrac{140}{-10}$ **6.** $\dfrac{-81}{3}$

Copyright © Big Ideas Learning, LLC
All rights reserved.

Lesson 12.4 Start Thinking!
For use before Lesson 12.4

A company's profits for a week are as follows: Monday: +\$32.65, Tuesday: –\$75.32, Wednesday: –\$125.75, Thursday: +\$100.89, and Friday: +\$65.30. Does the company show a gain or loss at the end of the week?

Lesson 12.4 Warm Up
For use before Lesson 12.4

Multiply. Write fractions in simplest form.

1. $-2\frac{1}{4} \bullet \frac{4}{5}$

2. $\left(-\frac{3}{5}\right)^3$

3. $-1.2(-3.05)$

Divide. Write fractions in simplest form.

4. $-2\frac{4}{7} \div (-3)$

5. $-9.6 \div 8$

6. $6.45 \div (-30)$

Copyright © Big Ideas Learning, LLC
All rights reserved.

Name______________________________ Date__________

12.4 Practice A

Tell whether the expression is *positive* or *negative* without evaluating.

1. $\dfrac{-7.5}{4.25}$
2. $\dfrac{4}{9} \times \left(-\dfrac{6}{7}\right)$
3. $-\dfrac{1}{5} \div \left(-\dfrac{2}{3}\right)$
4. $-3.2 \times (-1.7)$

Divide. Write fractions in simplest form.

5. $-\dfrac{2}{7} \div \dfrac{10}{7}$
6. $-\dfrac{1}{2} \div \left(-\dfrac{3}{4}\right)$
7. $\dfrac{2}{3} \div (-14)$
8. $-1\dfrac{1}{6} \div \dfrac{5}{3}$
9. $-0.72 \div (-0.9)$
10. $5.4 \div (-3.6)$

Multiply. Write fractions in simplest form.

11. $\dfrac{2}{5} \times \left(-\dfrac{10}{7}\right)$
12. $-\dfrac{3}{4} \bullet \left(-\dfrac{10}{9}\right)$
13. $\dfrac{3}{2}\left(-2\dfrac{2}{9}\right)$
14. $\left(-1\dfrac{3}{8}\right)^2$
15. -3.7×2.1
16. $-5.7 \bullet (-2.06)$

17. There are 15 people in a room. Each person ate $\dfrac{2}{3}$ of a pizza. There was no pizza remaining. How many pizzas were in the room?

18. During a drought, a river's height decreases by 0.35 inch every day. What is the change in the river's height after 7 days?

Evaluate.

19. $-3^2 + 4.6 \times (-0.1)$
20. $-2\dfrac{2}{3} \div 1\dfrac{5}{6} + 2$
21. $-4.31 \bullet 3.09 + (-0.98)$
22. $-3 \times \left(-1\dfrac{7}{12}\right) - \left(-\dfrac{3}{2}\right)^2$

23. Write two fractions, both not positive, whose product is $\dfrac{5}{8}$.

24. Fill in the blank to make the solution correct.

$$5.6 \times \underline{\ ?\ } = -19.04$$

Copyright © Big Ideas Learning, LLC
All rights reserved.

Name ______________________________ Date __________

12.4 Practice B

Divide. Write fractions in simplest form.

1. $-\frac{3}{7} \div \frac{11}{35}$

2. $-\frac{1}{9} \div \left(-\frac{13}{30}\right)$

3. $1\frac{5}{6} \div (-30)$

4. $-2\frac{4}{5} \div 10\frac{2}{3}$

5. $-0.801 \div (-0.09)$

6. $14.616 \div (-2.32)$

Multiply. Write fractions in simplest form.

7. $-\frac{2}{15} \times \left(-\frac{25}{6}\right)$

8. $-\frac{3}{14} \bullet \frac{21}{12}$

9. $1\frac{2}{3}\left(-2\frac{9}{10}\right)$

10. $-\left(3\frac{2}{5}\right)^2$

11. -2.75×3.1

12. $-1.27 \bullet (-2.02)$

13. How many three-quarter pound burgers can be made with twelve pounds of hamburger?

14. The table shows the changes in your times (in seconds) at the new skateboard ramp. What is your mean change?

Trial	1	2	3	4	5
Change	2.2	–1.4	0.6	–2.3	–1.7

Evaluate.

15. $-0.2^3 - 4.15(-0.06)$

16. $5 - 3\frac{9}{10} \div 2\frac{3}{5}$

17. $-14.01 \bullet 2.39 + |-4.89|$

18. $2\frac{1}{3} \times \left(-4\frac{5}{7}\right) - \left(-\frac{3}{5}\right)^2$

19. A gallon of gasoline costs \$2.96. Your car has a 25-gallon gas tank and can travel 28.8 miles on each gallon of gasoline.

a. Find the cost of filling your gas tank if it is already $\frac{3}{8}$ full.

b. You take a trip of length $705\frac{3}{5}$ miles. How much money do you spend on gasoline?

Copyright © Big Ideas Learning, LLC
All rights reserved.

Name ______________________________ Date __________

12.4 Enrichment and Extension

The Zweezam Factory

The Zweezam Factory manufactures Zweenubs, Zweedulls, and Zweebuds.

The table shows the costs and income for each type of item.

	Zweenubs	Zweedulls	Zweebuds
Cost per item manufactured	\$2.53	\$6.58	\$8.72
Income per item sold	\$4.50	\$8.89	\$9.99

Answer the following questions about this week at the Zweezam Factory.

1. This week the Zweezam Factory manufactured 46 items altogether including 24 Zweenubs. They manufactured two-thirds as many Zweedulls as Zweenubs and one-fourth as many Zweebuds as Zweenubs. How many of each type did the factory manufacture?

2. What is the factory's total manufacturing cost?

3. Also this week, the Zweezam Factory sold 32 parts altogether including 10 Zweedulls. They sold half as many Zweedulls as Zweenubs and one-fifth as many Zweebuds as Zweedulls. How many of each type did the factory sell?

4. What is the factory's total income?

5. What was the Zweezam Factory's profit this week? (*Hint:* profit = total income − total cost)

6. If the Zweezam Factory continues to manufacture and sell the same number of items for four more weeks, what will be the total profit for five weeks?

7. The Zweezam Factory plans to manufacture fewer items next week. They will manufacture 40 items altogether: two-fifths as many Zweedulls as Zweenubs and one-half as many Zweebuds as Zweedulls. If the Zweezam Factory manufactured 5 Zweebuds, how many of each type will the Zweezam Factory manufacture?

8. They already have pre-orders for next week's batch: 12 Zweenubs, 5 Zweedulls, and 1 Zweebud. Each week the parts must be thrown out, and only the new batch can be sold. So, how much more income do they have to make in order to break even (i.e. have a profit of zero) next week? What is the most profit they could earn next week?

9. If they sold Zweenubs only next week, could they make enough income to break even? Explain your reasoning.

Copyright © Big Ideas Learning, LLC
All rights reserved.

Name ______________________________ Date __________

12.4 Puzzle Time

When Is A Baby Like A Basketball Player?

Write the letter of each answer in the box containing the exercise number.

Multiply. Write fractions in simplest form.

1. $-\frac{4}{5} \bullet \left(-\frac{5}{7}\right)$

2. $2\frac{2}{3} \bullet \left(-4\frac{1}{4}\right)$

3. $\left(-\frac{3}{4}\right)^3$

4. $0.8 \times (-2.1)$

5. $-7.5 \times (-0.3)$

6. $(-0.8)^3$

Divide. Write fractions in simplest form.

7. $\frac{5}{8} \div \left(-\frac{1}{4}\right)$

8. $-1\frac{1}{6} \div \frac{2}{9}$

9. $-6\frac{2}{5} \div \left(-2\frac{2}{7}\right)$

10. $0.3 \div (-1.5)$

11. $-5.415 \div (-2.85)$

12. $-16.29 \div 3.62$

13. What is the square foot area of a room with a length of $10\frac{3}{4}$ feet and a width of $8\frac{1}{2}$ feet?

14. For a fundraiser, the seventh grade class sells 45 submarine sandwiches. They collect a total of $150.75. What is the cost per sub?

Answers

R. 2.25	**E.** $-\frac{27}{64}$
S. $-2\frac{1}{2}$	**D.** $91\frac{3}{8}$
H. -0.512	**E.** $-5\frac{1}{4}$
B. 3.35	**I.** -0.2
L. -4.5	**W.** 1.9
E. $\frac{4}{7}$	**B.** -1.68
N. $2\frac{4}{5}$	**H.** $-11\frac{1}{3}$

11	6	3	9		2	8		13	5	10	14	4	12	1	7

Copyright © Big Ideas Learning, LLC
All rights reserved.

Name____ Date __________

Chapter 12

Technology Connection

For use after Section 12.3

Adding and Subtracting Rational Numbers

In this activity, you will use a scientific calculator to add and subtract rational numbers. To enter a fraction on a scientific calculator, use the [a b/c] key.

To convert a mixed number to an improper fraction, use the [d/c] key.

EXAMPLE Use a scientific calculator to find the sum or difference.

a. $\frac{1}{4} + \frac{2}{3}$ **b.** $-\frac{21}{8} - \frac{17}{12}$ **c.** $6\frac{3}{10} + \left(-2\frac{5}{8}\right)$

SOLUTION

Enter the following keystrokes:

a. 1 [a b/c] 4 [+] 2 [a b/c] 3 [=]

The answer is $\frac{11}{12}$.

b. 21 [+/−] [a b/c] 8 [−] 17 [a b/c] 12 [=]

The answer is $-4\frac{1}{24}$.

c. 6 [a b/c] 3 [a b/c] 10 [+] 2 [+/−] [a b/c] 5 [a b/c] 8 [=]

The answer is $3\frac{27}{40}$.

Notice in parts (b) and (c) that the final display shows a mixed number. To convert the mixed number in part (b) to an improper fraction, press [2nd] or [Shift] then [d/c]. The answer as an improper fraction is $-\frac{97}{24}$.

Use a scientific calculator to find the sum or difference. Write your answer as a mixed number and an improper fraction, if possible.

1. $\frac{1}{5} + \frac{7}{10}$

2. $\frac{11}{4} - \frac{3}{8}$

3. $-\frac{4}{9} - \left(-\frac{1}{12}\right)$

4. $\frac{14}{25} + \frac{38}{15}$

5. $3\frac{5}{16} + \left(-2\frac{1}{2}\right)$

6. $-5\frac{1}{4} + 6\frac{5}{9}$

Copyright © Big Ideas Learning, LLC
All rights reserved.

Chapter 13

Copyright © Big Ideas Learning, LLC
All rights reserved.

Name____ Date __________

Chapter 13 Expressions and Equations

Dear Family,

Algebra is used to describe relationships in general terms. Consider the following statements.

• Game tickets are $7 each.	The cost of n tickets is $7n$ dollars.
• It takes 5 minutes to get shoes and car keys and walk to the car.	For a drive of m minutes, allow $m + 5$ minutes.
• Each question on a 20-question test is worth 1 point.	If you miss x questions, your score on the test will be $20 - x$.

On the left, the rule is stated in words, the way you might remember it. On the right, the rule is stated as a mathematical expression with a variable. The number of tickets, the length of the drive, and the number of questions missed are all variables—that is, they might have many different values. The cost of a ticket, the time to get to the car, and the total number of questions on the test are constants—that is, they remain the same. Ask your student to answer each question, using the information above.

- What is the cost of 3 game tickets?
- You want to arrive at baseball practice at 4:30. The drive is 15 minutes. What time should you get ready to leave?
- You miss 2 questions on the test. What is your score?

(Answers: $21, 4:10, 18 points)

Rather than remember all possible ticket costs, driving times, or test scores, you remember the rule for finding them. These examples are uses of algebra in daily life.

With your student, find another algebraic rule you could use in daily life. What are the variables? What are the constants? Have your student evaluate your rule for two different values of the variable(s).

Have fun exploring expressions together!

Copyright © Big Ideas Learning, LLC
All rights reserved.

Nombre ______________________________ Fecha ________

Capítulo 13 Expresiones y ecuaciones

Estimada Familia:

El álgebra se utiliza para describir relaciones en términos generales. Consideren los siguientes enunciados.

- Los boletos para el juego cuestan \$7 cada uno. — n boletos cuestan $7n$ dólares.
- Se necesitan 5 minutos para ponerse los zapatos, tomar las llaves del auto y caminar hacia él. — Para un trayecto de m minutos, considere $m + 5$ minutos.
- Cada pregunta en una prueba de 20 preguntas vale 1 punto. — Si falla x respuestas, su puntaje en la prueba será $20 - x$.

A la izquierda, la regla se expresa en palabras, la manera en que podrían recordarla. A la derecha, la regla se indica en una expresión matemática con una variable. El número de boletos, la duración del trayecto y el número de preguntas erróneas son todas variables—es decir, que pueden tener muchos valores diferentes. El costo de un boleto, el tiempo para llegar al auto y el número total de preguntas en la prueba son constantes—es decir, siguen siendo los mismos. Pida a su estudiante que responda cada pregunta utilizando la información anterior.

- ¿Cuánto cuestan 3 boletos para el juego?
- Desean llegar al entrenamiento de béisbol a las 4:30. El trayecto dura 15 minutos. ¿A qué hora deben alistarse para salir?
- Se equivoca en 2 preguntas en la prueba. ¿Cuál es su puntaje?

(Respuestas: \$21, 4:10, 18 puntos). En lugar de recordar todos los gastos posibles de boletos, los tiempos de manejo o los resultados de la prueba, debe recordar la regla para hallarlos. Estos ejemplos son usos del álgebra en la vida diaria.

Con su estudiante, busquen otra regla algebraica que se pueda utilizar en la vida diaria. ¿Cuáles son las variables? ¿Cuáles son las constantes? Haga que su estudiante evalúe la regla para dos valores diferentes de la(s) variable(s).

Copyright © Big Ideas Learning, LLC
All rights reserved.

Activity 13.1 Start Thinking!

For use before Activity 13.1

Explain to a partner how an expression and an equation are different.

Give an algebraic and numerical example for each.

Activity 13.1 Warm Up

For use before Activity 13.1

Write the phrase as an expression.

1. 7 increased by a number x

2. negative 14 minus y

3. negative 19 increased by n

4. the product of 14 and y

5. 10 divided by the sum of a number n and 6

6. 6 times the quotient of a number x and 3

Copyright © Big Ideas Learning, LLC
All rights reserved.

Lesson 13.1

Start Thinking!

For use before Lesson 13.1

Given the problem $6x + 4 - 2x$, your brother says the answer is $8x + 4$. Explain to your brother why his answer is incorrect. Give the correct answer.

Lesson 13.1

Warm Up

For use before Lesson 13.1

Simplify the expression.

1. $10x + 4x$

2. $7y + 12 - 15$

3. $6x - 4x + 16$

4. $3.8y - 4 + 7.2y$

5. $7 + 11x + 8.4 - x$

6. $\frac{4}{7}y + 8 - 2\frac{1}{2} + \frac{3}{7}y$

Copyright © Big Ideas Learning, LLC
All rights reserved.

Name________________________________ Date __________

13.1 Practice A

Identify the terms and like terms in the expression.

1. $-4y + 7 + 9y - 3$

2. $3n^2 - 1.4n + 5n^2 - 6.4$

3. $\frac{1}{2}b^3 - b^3 + 2b$

Simplify the expression.

4. $-15m + 9m$

5. $8k - 2(4 - 3k)$

6. $3.2 - 9x + 7.1 - 3x$

7. $25 - 6x - 12 - 2x$

8. $19a - 7 - 3a + 12a$

9. $\frac{5}{2}(6x - 7) + \frac{4}{3}(2 + 9x)$

10. $\frac{1}{8}h + 7 - \frac{3}{4}h$

11. $\frac{2}{3}y + 5 - 3 - \frac{11}{12}y$

12. Write an expression in simplest form that represents the perimeter of the polygon.

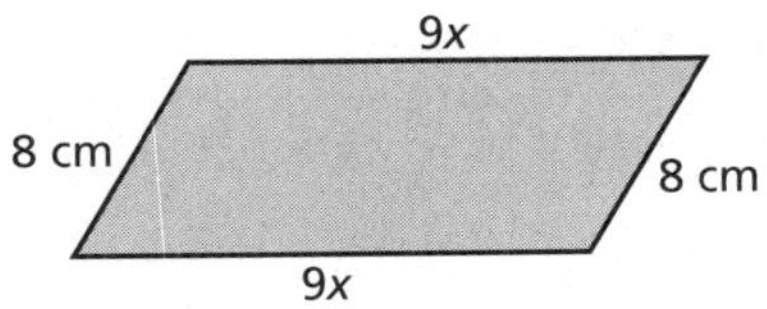

13. Each runner is carrying an 8 ounce bottle of water, a 2.1 ounce energy bar, and a 3 ounce energy drink. Write an expression in simplest form that represents the weight carried by y runners. Interpret the expression.

14. John weighs 65 kilograms, Sam weighs $22x$ kilograms, and Mark weighs $13x$ kilograms. Write an expression in simplest form for their combined weight.

15. Are the expressions $8a^2 - 4b + 7a^2$ and $5(3a^2 - 2b) + 6b$ equivalent? Explain your reasoning.

Copyright © Big Ideas Learning, LLC
All rights reserved.

Name ______________________________ Date __________

13.1 Practice B

Identify the terms and like terms in the expression.

1. $1.3x - 2.7x^2 - 5.4x + 3$

2. $10 - \frac{3}{10}m + 6m^2 + \frac{2}{5}m$

Simplify the expression.

3. $-\frac{15}{4}b + \frac{5}{6}b$

4. $60m - 15(4 - 8m) + 20$

5. $4(5.8 - 9x) + 8.2 + 22x$

6. $9y - 15y + 12 - 6y$

7. $v + 13 - 8(v + 2)$

8. $\frac{5}{3}(5x + 9) + \frac{4}{5}(1 - 9x)$

9. Write an expression in simplest form that represents the perimeter of the polygon.

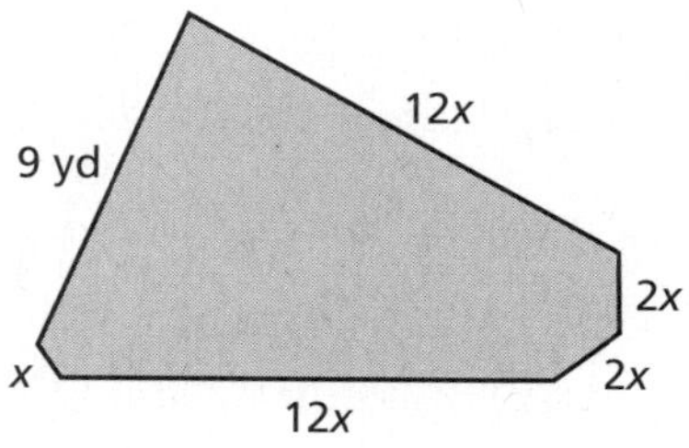

Draw a diagram that shows how the expression can represent the area of a figure. Then simplify the expression.

10. $8(3x - 1)$

11. $(5 + 2)(x + 3x)$

12. Danielle is x years old. Her sister is 5 years older and her brother is half Danielle's age. Write an expression in simplest form for the sum of their ages.

13. The length of a rectangular field is 30 more than twice its width. Write an expression in simplest form for the perimeter of the field in terms of its width w.

14. You buy x packs of pencils, twice as many packs of erasers, and three times as many rolls of tape. Write an expression in simplest form for the total amount of money you spent.

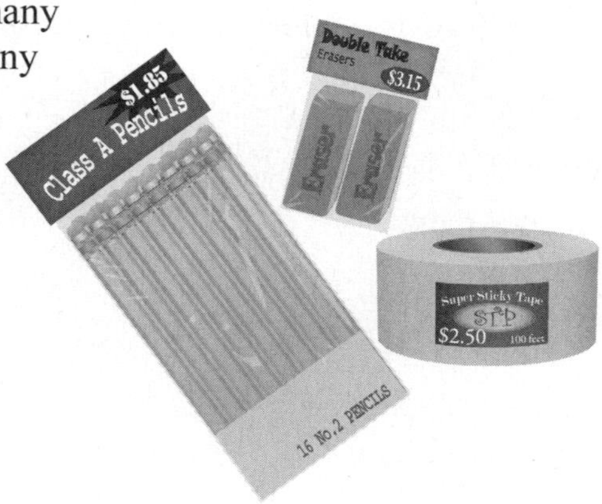

Copyright © Big Ideas Learning, LLC
All rights reserved.

Name____ Date__________

13.1 Enrichment and Extension

Matching

Simplify the expressions on the left by using the Distributive Property and combining like terms. Then, match it to an equal expression on the right by connecting the two with a line.

1. $6x + 2x$	**a.** $8x$
2. $14x - 12 - x - 3$	**b.** $\frac{1}{2}x + 1$
3. $-5x + 14 - x - 2$	**c.** $13x - 15$
4. $-3 - 5x - 3x + 11x + 3$	**d.** $2x + 11$
5. $-2(-5 - x) + x - x + 1$	**e.** $2x$
6. $\frac{1}{2}(12) + 4x - (x - 1)$	**f.** $6x^2 + x - 27$
7. $6(x^2 - 2) + 1 - 16 + x$	**g.** $3x$
8. $4\left(\frac{1}{2}x + 4\right) + 1 - 16 + x$	**h.** $3x + 1$
9. $5(x^2 + x)$	**i.** $3x + 7$
10. $x + \left(1 - \frac{1}{2}x\right)$	**j.** $-6x + 12$
11. $x^3 + x^2 + x + x - x^2 - x^3$	**k.** $5x^2 + 5x$

12. Write an expression containing x-terms and constants. The x-terms should combine to $7x$ and the constants should sum to 13.

13. Write an expression containing x^2-terms, x-terms, and constants. The x^2-terms should combine to $-2x^2$, the x-terms should subtract to $3x$, and the constants should sum to 3.

Copyright © Big Ideas Learning, LLC
All rights reserved.

Name ______________________________ Date ________

13.1 Puzzle Time

How Can You Turn A Pumpkin Into A Squash?

A	B	C	D	E	F
G	H	I	J	K	L

Complete each exercise. Find the answer in the answer column. Write the word under the answer in the box containing the exercise letter.

$2x + 4$	SMASH
$13x - 2$	THE
$-2x + 6.2$	COME
$2.4x + 2.9$	AND
$21x$	THROW
$-1.5x - 7$	WILL
$7x + 14$	SQUASH

Simplify the expression.

A. $8x + 13x$

B. $15x + 10 - 6$

C. $7x - 4x + 3$

D. $5.3x - 9 + 7.6x$

E. $6x - 4x - 2 + 11x$

F. $\frac{3}{4}x + 11 - 5\frac{1}{2} + \frac{1}{4}x$

G. $5(x + 8) + 3$

H. $3.6x - 7 - 5.1x$

I. $4 + 8x + 2.2 - 10x$

J. $\frac{5}{6}x - 9 + 3 - \frac{2}{3}x$

K. $2.4(x + 3) - 4.3$

L. The length of a rectangle is 7 inches and the width is $(x + 2)$ inches. Write an expression in simplest form that represents the area of a rectangle.

$x + 5\frac{1}{2}$	AIR
$3x + 3$	UP
$5x + 43$	IT
$x - 4\frac{1}{2}$	TOSS
$12.9x - 9$	IN
$\frac{1}{6}x - 6$	DOWN
$15x + 4$	IT

Copyright © Big Ideas Learning, LLC
All rights reserved.

Activity 13.2 Start Thinking!

For use before Activity 13.2

Give two examples of *like terms*. Give two examples of *unlike terms*. Explain what the difference is between like terms and unlike terms.

Activity 13.2 Warm Up

For use before Activity 13.2

Simplify the expression.

1. $4x + 2 - 3x$

2. $8y - 3 - 10y - 6$

3. $-2x + 3 - 8x$

4. $\frac{1}{2}y + 3y - \frac{2}{5}$

5. $4.8x - 4.6 + 3.9x$

6. $3y - 4.6 + 1.3 + 2.1y$

Copyright © Big Ideas Learning, LLC
All rights reserved.

Lesson 13.2 Start Thinking!

For use before Lesson 13.2

Explain how to simplify $(5x - 4) + (3x - 6)$ using algebra tiles.

Lesson 13.2 Warm Up

For use before Lesson 13.2

Write the sum or difference of two algebraic expressions modeled by the algebra tiles. Then use the algebra tiles to simplify the expression.

1. 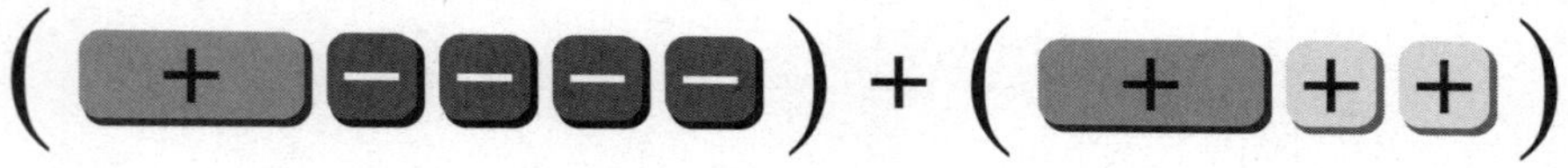

2. 

3. 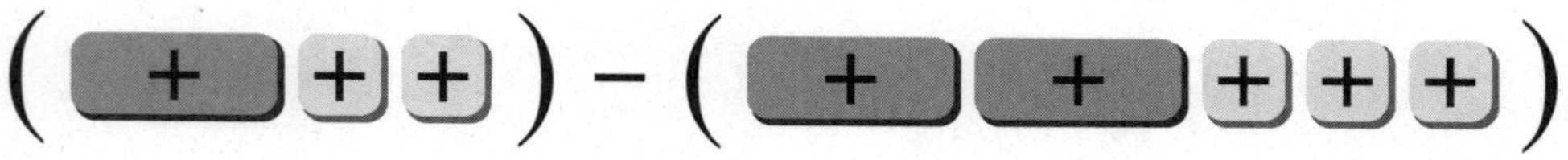

4. 

Copyright © Big Ideas Learning, LLC
All rights reserved.

Name______________________________ Date__________

13.2 Practice A

Find the sum.

1. $(p - 3) + (p - 7)$

2. $(3n - 1) + (4 - n)$

3. $(-3r + 8) + (5r - 1)$

4. $6(x - 3) + (2x - 9)$

5. $(3c + 2) + 4(1.3c - 5)$

6. $10(2.1q - 2) + (7.5q + 18)$

7. $(-6y - 2) + 5(3 + 2.5y)$

8. $\frac{1}{2}(6x - 10) + \frac{1}{3}(6 + 9x)$

9. After a week of rain, tadpoles appeared in your pond. After t minutes, you have $(7t + 5)$ tadpoles and your friend has $(8t - 3)$ tadpoles.

   a. Write an expression that represents the number of tadpoles you and your friend caught together.

   b. Who has more tadpoles after 9 minutes?

Find the difference.

10. $(k + 3) - (3k - 5)$

11. $(-6d + 2) - (7 + 2d)$

12. $(10j - 7) - (-9j + 2)$

13. $(3x + 8) - 6(2.5x - 3)$

14. $(7 - 3t) - 5(-1.6t + 5)$

15. $\frac{1}{2}(12w + 8) - \frac{1}{5}(10w - 5)$

16. The admission to a local fair is \$10 for each adult and \$6 for each child. Each rides costs \$1.50 for an adult and \$1 for a child.

    a. Write an expression that represents how much more an adult will spend at the fair.

    b. An adult and a child each go on 7 rides. How much more did the adult spend?

17. Write an expression that represents the perimeter of the triangle.

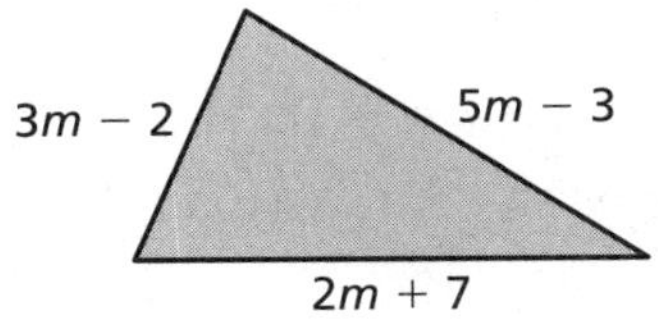

Copyright © Big Ideas Learning, LLC
All rights reserved.

Name ______________________________ Date ________

13.2 Practice B

Find the sum.

1. $(5 - t) + (3t + 2)$
2. $(7k + 9) + (4k - 3)$
3. $2(-5y + 6) + (2y - 8)$
4. $4(2.5g - 4) + 3(1.2g - 2)$
5. $5(-0.3s - 2) + 2(5 - 3.4s)$
6. $\frac{1}{3}(6p - 3) + \frac{1}{7}(7p + 14)$
7. $\frac{1}{5}(-15w - 20) + \frac{1}{2}(3 - 4w)$
8. $\frac{1}{8}(16k - 24) + \frac{1}{5}(2 + 10k)$
9. You are selling tickets to a play. You have sold $(3t + 2)$ tickets for \$5 each and $(2t + 5)$ tickets for \$7 each.
   a. Write an expression that represents the total number of tickets sold so far.
   b. Write an expression that represents the total amount of money received for the tickets that have been sold.
   c. When $t = 3$, what is the total amount of money received?

Find the difference.

10. $(8 - u) - (5u + 1)$
11. $(2x + 7) - 6(5x - 8)$
12. $(3h + 4) - 6(5 - 1.4h)$
13. $(12 + 7.2b) - 3(0.9b - 4)$
14. $\frac{1}{4}(16j - 12) - \frac{1}{9}(18j + 45)$
15. $\frac{2}{3}(12n + 6) - \frac{1}{5}(10n - 2)$
16. You are collecting pairs of socks and toothbrushes for a local charity. After *d* days, you have collected $(4d + 5)$ pairs of socks and $(3d + 7)$ toothbrushes.
    a. Write an expression that represents the total number of items that have been collected.
    b. How many more pairs of socks than toothbrushes have been collected on day 7?

Copyright © Big Ideas Learning, LLC
All rights reserved.

Name____ Date__________

13.2 Enrichment and Extension

Using the Distributive Property

When working with algebraic expressions and the Distributive Property, the exponents of the variables are added.

Example: Simplify $x(x + 6)$.

Distribute x to each term inside the parentheses. (Remember that x can be rewritten as $1 \bullet x^1$.) Then, multiply the coefficients.

$x(x + 6) = (1x \bullet 1x) + (1x \bullet 6)$	Distibute x to each term.
$= (1x^1 \bullet 1x^1) + (1x^1 \bullet 6)$	Rewrite to show exponents.
$= 1x^{1+1} + 6$	Multiply coefficients and add exponents.
$= x^2 + 6x$	Simplify.

Simplify the expression.

1. $x^2(x + 1)$
2. $-x(2x - 8)$
3. $x(x^4 - 4)$
4. $x(3x - 1)$
5. $3x(x - 1)$
6. $2x(x - 1)$
7. $4x(-4x - 3)$
8. $n(n - 4)$
9. $-b(3b + 9)$
10. $2w(-4w - 14)$
11. $2x(4x - 9) - 3x(4x - 2)$
12. $3k(-5k + 21) + 2(2.5k + 9)$
13. $4(1 + 1.8h) + h(2.2h + 5)$
14. $\frac{1}{2}m(10 + 6m) - \frac{1}{5}m(10m + 10)$
15. $\frac{1}{3}(6z - 6) - \frac{1}{4}z(4z + 16)$
16. $3d^6(24d^3 + 6) + 24d^9$

Copyright © Big Ideas Learning, LLC
All rights reserved.

Name ______________________________ Date __________

13.2 Puzzle Time

What Did The Candle Say To The Match?

Write the letter of each answer in the box containing the exercise number.

Find the sum.

1. $(x + 10) + (x - 14)$

2. $(9 - 2x) + (6x + 4)$

3. $(3x - 7) + (-4x - 8)$

4. $(2x - 7) + 5(x - 3)$

5. $6(-2.3x - 5) + (4x + 11)$

6. $(8 - 2x) + 3(4.5x + 9)$

7. $\frac{1}{2}(8 - 4x) + \frac{1}{3}(9x - 6)$

8. $-\frac{3}{4}(3x + 7) + \frac{1}{4}(12x + 20)$

Find the difference.

9. $(-3x + 8) - (x + 10)$

10. $(5x + 4) - (1 - 2x)$

11. $(3 - 4x) - 3(2.4x - 7)$

12. $(4x - 8) - 4(-6.5x + 5)$

13. $\frac{1}{9}(-9x + 18) - \frac{1}{5}(10 + 15x)$

14. $\frac{4}{7}(4x + 3) - \frac{1}{7}(9x + 5)$

15. $\frac{1}{2}(-4x + 8) - \frac{1}{4}(8x - 12)$

16. Your class project involves recycling aluminum cans. After x weeks, your class has $(13x + 50)$ aluminum cans. The class goal is to collect $(80x + 120)$ aluminum cans. How many more aluminum cans does your class need to collect?

Answers

U. $-4x - 2$

P. $30x - 28$

T. $-9.8x - 19$

E. $x + 2$

I. $2x - 4$

L. $67x + 70$

H. $-11.2x + 24$

Y. $7x - 22$

I. $4x + 13$

U. $\frac{3}{4}x - \frac{1}{4}$

G. $x + 1$

L. $-4x + 7$

Y. $11.5x + 35$

F. $-4x$

M. $7x + 3$

O. $-x - 15$

6	3	9		16	2	14	11	5		8	12		10	4		15	1	13	7

Copyright © Big Ideas Learning, LLC
All rights reserved.

Extension 13.2

Start Thinking!

For use before Extension 13.2

Your friend says the greatest common factor of 15 and 30 is 5. Is your friend correct? Explain your reasoning.

Extension 13.2

Warm Up

For use before Extension 13.2

Find the GCF.

1. 2, 10

2. 3, 24

3. 9, 27

4. 14, 49

5. 24, 42

6. 50, 90

Copyright © Big Ideas Learning, LLC
All rights reserved.

Name ______________________________ Date __________

Extension 13.2 Practice

Factor the expression using the GCF.

1. $8 - 22$ **2.** $25 + 30$ **3.** $6y + 3$

4. $2t - 10$ **5.** $16p - 8$ **6.** $21s + 15$

7. $32v + 24w$ **8.** $9b + 24c$ **9.** $12y - 42z$

Factor out the coefficient of the variable.

10. $\frac{1}{2}m + \frac{1}{2}$ **11.** $\frac{2}{3}j - \frac{2}{9}$ **12.** $1.2k + 2.4$

13. $1.5a - 4.5$ **14.** $3f + 5$ **15.** $\frac{3}{10}x - \frac{3}{5}$

16. Factor $-\frac{1}{3}$ out of $-\frac{1}{3}x - 12$.

17. Factor $-\frac{1}{6}$ out of $-\frac{1}{3}x + \frac{5}{6}y$.

18. The area of the rectangle is $(18x - 12)$ square inches. Write an expression that represents the length of the rectangle (in inches).

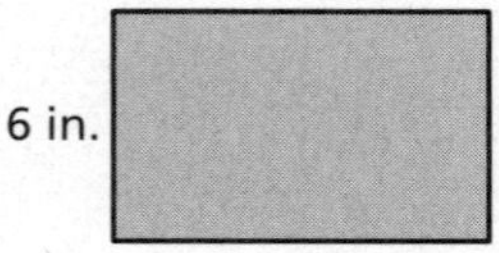

19. A concession stand sells hamburgers. The revenue from the hamburgers is $(30x + 45)$ dollars.

a. The price of a hamburger is \$5. Write an expression that represents the number of hamburgers sold.

b. The revenue from drinks is $(63x + 84)$ dollars. The price of a drink is \$3. Write an expression that represents the number of drinks sold.

c. Write and simplify an expression that represents how many more drinks were sold.

Copyright © Big Ideas Learning, LLC
All rights reserved.

Activity 13.3 Start Thinking!

For use before Activity 13.3

You have 7 less points than your cousin. Your brother has 8 more points than your sister.

Write an expression to model each situation. Use p as your variable. Can each expression be written in more than one way? Explain.

Activity 13.3 Warm Up

For use before Activity 13.3

Add.

1. $65 + (-23)$ **2.** $-12 + (-34)$ **3.** $-35 + 42$

Subtract.

4. $-15 - 24$ **5.** $29 - 35$ **6.** $52 - (-13)$

Copyright © Big Ideas Learning, LLC
All rights reserved.

Lesson 13.3

Start Thinking!

For use before Lesson 13.3

Discuss with a partner, using an example, how inverse operations are used to solve equations.

Lesson 13.3

Warm Up

For use before Lesson 13.3

Solve the equation. Check your solution.

1. $x + 5 = 10$

2. $y - 2 = 16$

3. $n - 13 = 65$

4. $18 = p + 3$

5. $34 = t - 23$

6. $z + 14 = 21$

Copyright © Big Ideas Learning, LLC
All rights reserved.

Name____ Date__________

13.3 Practice A

Solve the equation. Check your solution.

1. $x + 3 = 10$

2. $b - 6 = -14$

3. $5 = n + 9$

4. $y - 2.1 = 7.5$

5. $-6.4 = x + 4.3$

6. $k - \frac{1}{3} = \frac{5}{6}$

7. $10.5 + p = -8.32$

8. $3\frac{3}{4} = r + \frac{1}{8}$

9. $m + 1.06 = 5$

10. $-\frac{7}{12} = 1\frac{5}{6} + d$

11. $t - \frac{2}{7} = \frac{1}{2}$

12. $-10.2 + c = -8.14$

Write the word sentence as an equation. Then solve.

13. 5 more than a number y is -2.

14. The sum of 8 and a number h is 12.

15. -13 is 4 less than a number n.

In Exercises 16–20, write an equation. Then solve.

16. You earn $9 per hour babysitting. This is $2 more than what you earned per hour last year. What did you earn per hour last year?

17. Your mother asked you to turn the oven down to 325°F. This is 75°F less than it was. What was the original temperature?

18. The difference between the heights of your chair and your desk is $-10\frac{1}{4}$ inches. The height of your desk is $29\frac{3}{4}$ inches. What is the height of your chair?

19. Your Two-Day-Pass to a theme park is $76.50. This is $31.41 less than your uncle's Two-Day-Pass. What is the price of your uncle's pass?

20. The perimeter of a triangle is 25 feet. What is the length of the unknown side?

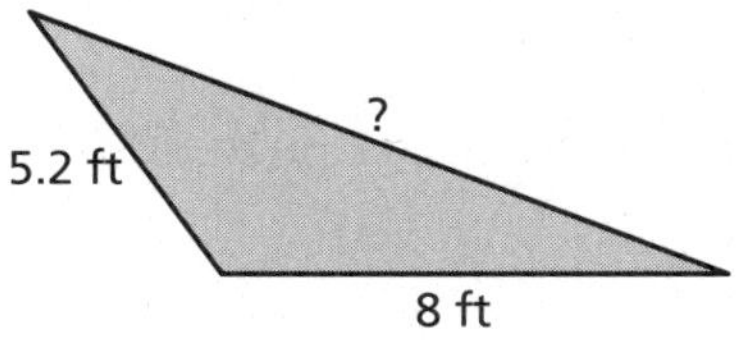

21. Find the value of $3x + 2$ when $7 + x = 5$.

Copyright © Big Ideas Learning, LLC
All rights reserved.

Name ______________________________ Date __________

13.3 Practice B

Solve the equation. Check your solution.

1. $x + 12 = 100$

2. $g - 16 = -52$

3. $-4.5 = m + 1.9$

4. $j - 12.1 = 7.53$

5. $y + 4.003 = -3.14$

6. $z - 4\frac{1}{3} = 2\frac{2}{3}$

7. $90.8 + q = -18.24$

8. $10\frac{2}{3} = r + 12\frac{1}{6}$

9. $b + 4.006 = 9$

10. $-7\frac{5}{8} = 1\frac{5}{6} + d$

11. $f - \frac{2}{15} = 6\frac{3}{5}$

12. $-10.216 + c = -12.014$

Write the word sentence as an equation. Then solve.

13. 27 is 12 more than a number x.

14. The difference of a number p and -9 is 12.

15. 35 less than a number m is -72.

In Exercises 16–18, write an equation. Then solve.

16. You swim the 50-meter freestyle in 28.12 seconds. This is 0.14 second less than your previous fastest time. What was your previous fastest time?

17. The perimeter of a rectangular backyard is $32\frac{1}{2}$ meters. The two shorter sides are each $7\frac{3}{8}$ meters long. What is the length of the two longer sides? (*Hint*: The sum of the shorter side and the longer side is equal to half of the perimeter.)

18. The temperature of dry ice is $-109.3°F$, which is $183.6°F$ less than the outside temperature. What is the outside temperature?

19. Your cell phone bill in August was $61.43, which was $21.75 more than your bill in July. Your cell phone bill in July was $13.62 less than your bill in June. What was your cell phone bill in June?

Find the values of *x*.

20. $|x| - 10.5 = 4.3$

21. $|x + 2| - 7 = 5$

Copyright © Big Ideas Learning, LLC
All rights reserved.

Name____ Date__________

13.3 Enrichment and Extension

You Be the Teacher

In Exercises 1 and 2, use the student solutions below.

Courtney	Karen
$\lvert x \rvert = 7$	$\lvert x \rvert = 7$
$x = 7$	$x = 7$ or $x = -7$

1. Did both students get a correct solution?

2. Is one student's answer more complete? If so, which one? Explain.

3. A student asks you how to solve $\lvert x + 8 \rvert = 12$. Describe your explanation and any math steps you would show.

4. Describe Mario's solution. Did he get the correct answers? Explain.

Mario

$\lvert x - 5 \rvert = 9$

$x - 5 = 9$ or $x - 5 = -9$

$x = 14$ or $x = -4$

5. Kelly looks at Mario's solutions and does not understand why -4 is a solution. She thought absolute value could not be negative. Explain Kelly's error.

6. Pat says the solutions of $\lvert x \rvert + 7 = 2$ are $x = 5$ and $x = -5$. What mistake did Pat make? Explain.

7. A student asks you if every absolute value equation has two solutions. How would you respond? Explain.

8. Give an example of an absolute value equation with (a) one solution, (b) two solutions, and (c) no solutions.

Copyright © Big Ideas Learning, LLC
All rights reserved.

Name ______________________ Date __________

13.3 Puzzle Time

What Did The Digital Clock Say To Its Mother?

Circle the letter of each correct answer in the boxes below. The circled letters will spell out the answer to the riddle.

Solve the equation.

1. $x + 8 = 21$

2. $3 = a - 12$

3. $y - 7 = -4$

4. $g + 11 = -13$

5. $z - 1.75 = 3.82$

6. $4.9 = h - 2.6$

7. $8.7 + b = 14.5$

8. $-10.3 = w - 5.8$

9. $\frac{3}{5} = c + \frac{1}{4}$

10. $r + 3\frac{1}{2} = -4\frac{2}{3}$

11. $5\frac{3}{4} = d - 2\frac{1}{8}$

12. $-7\frac{1}{3} = p - \frac{4}{9}$

13. The second book in your favorite series has 9 more chapters than the first book in the series. The second book has 38 chapters. How many chapters does the first book have?

14. Emily has a Springer Spaniel that weighs 48.5 pounds. She also has a Cocker Spaniel that weighs 24.8 pounds less than the Springer Spaniel. How many pounds does the Cocker Spaniel weigh?

L	M	O	O	I	K	E	T	M	S	O	R	D	M	A	E
23.7	4.7	3	$-8\frac{1}{6}$	$2\frac{3}{7}$	5.8	$-5\frac{2}{5}$	7.2	13	−42	−4.5	$1\frac{1}{9}$	−8.9	$-6\frac{8}{9}$	12	8
H	**N**	**E**	**R**	**O**	**S**	**H**	**T**	**U**	**A**	**N**	**Y**	**D**	**M**	**E**	**S**
52	7.5	$4\frac{3}{8}$	$12\frac{1}{3}$	−24	$-1\frac{2}{5}$	29	−3.9	17	5.57	$\frac{7}{20}$	$-\frac{1}{6}$	15	33	65.5	$7\frac{7}{8}$

Copyright © Big Ideas Learning, LLC
All rights reserved.

Activity 13.4 Start Thinking!

For use before Activity 13.4

Explain how buying a number of items at a store is like solving a math problem.

Activity 13.4 Warm Up

For use before Activity 13.4

Multiply.

1. $(-21)(8)$

2. $(-18) \bullet (-12)$

3. $5 \bullet (-13)$

Divide.

4. $\dfrac{-96}{3}$

5. $\dfrac{108}{-12}$

6. $\dfrac{-128}{-8}$

Copyright © Big Ideas Learning, LLC
All rights reserved.

Lesson 13.4 Start Thinking!

For use before Lesson 13.4

With a partner, write and solve a real-life word problem using the equation $13x = 39$.

Then rewrite the word problem using division.

Lesson 13.4 Warm Up

For use before Lesson 13.4

Solve the equation. Check your solution.

1. $4x = 24$

2. $2x = 56$

3. $-7x = 77$

4. $\frac{x}{5} = 12$

5. $\frac{x}{7} = 9$

6. $\frac{x}{-8} = 6$

Copyright © Big Ideas Learning, LLC
All rights reserved.

Name____ Date__________

13.4 Practice A

Solve the equation. Check your solution.

1. $4b = 24$

2. $-7n = 35$

3. $\frac{y}{-3} = 33$

4. $\frac{p}{5} = -32$

5. $-3t = -4.2$

6. $1.5q = -8.4$

7. $\frac{1}{5}d = -3$

8. $14 = 3y$

9. $\frac{g}{2.1} = -6.8$

10. $-\frac{3}{5}a = 2$

11. $\frac{k}{-9} = -\frac{1}{3}$

12. $\frac{5}{8}j = -10$

Write the word sentence as an equation. Then solve.

13. A number multiplied by $\frac{1}{2}$ is $-\frac{5}{12}$.

14. The quotient of a number and 0.2 is -2.6.

In Exercises 15–19, write an equation. Then solve.

15. You earn \$7.50 per hour at a fast food restaurant. You earned \$123.75 last week. How many hours did you work last week?

16. Your family took a road trip on Saturday. You were in the car for 4.5 hours and averaged 70 miles per hour. How many miles did you travel?

17. The area of a rectangle is $\frac{1}{2}$ square inch. The length of the rectangle is $\frac{3}{8}$ inch. What is the width of the rectangle?

18. You are in a room with other students and are asked to get in groups of 3. When finished, there are 21 groups of 3. How many students are in the room?

19. The perimeter of a square is 26.46 inches. What is the side length of the square?

20. Write a multiplication equation that has a solution of $\frac{2}{7}$.

21. Write a division equation that has a solution of -20.

Copyright © Big Ideas Learning, LLC
All rights reserved.

Name ______________________________ Date __________

13.4 Practice B

Solve the equation. Check your solution.

1. $16t = 60$

2. $-14p = -21$

3. $\frac{q}{5} = -7.35$

4. $\frac{d}{1.2} = -3.3$

5. $-\frac{8}{15}k = -4$

6. $-7.24q = 17.014$

7. $\frac{1}{8}d = -\frac{3}{5}$

8. $1.5 = 3.3y$

9. $\frac{g}{0.003} = -2.8$

10. $-\frac{10}{21}c = -\frac{15}{28}$

11. $\frac{k}{-9} = -1$

12. $18 = -\frac{6}{11}h$

In Exercises 13 and 14, write an equation. Then solve.

13. You order an entree for \$12.00. You pay \$0.78 in taxes. What is the tax rate?

14. If a project is handed in late, you receive $\frac{8}{9}$ of your earned points. You received 72 points on your late project. How many points did you lose?

15. Write a multiplication equation that has a solution of -14.8.

16. Write a division equation that has a solution of $-\frac{9}{14}$.

17. There are 92 students in a room. They are separated into 18 groups. How many students are in each group? How many students are not in a group?

18. A bus token costs \$1.75.

a. You spend \$15.75 on tokens. Write and solve an equation to find how many tokens you purchase.

b. If you purchase 10 tokens, you get 2 free tokens. Write and solve an equation to find the approximate reduced price of each token.

c. You also receive free tokens if you purchase 20 tokens. The reduced price for each token is \$1.40. Write and solve an equation to find how many free tokens you receive.

19. Solve $\frac{1}{3}|z| = 2$.

Copyright © Big Ideas Learning, LLC
All rights reserved.

Name____ Date__________

13.4 Enrichment and Extension

Equations with No Solution

For some equations, there is no value that could be substituted for the variable to make the equation true. In this case, the equation has no solution.

Example: Solve $4|x| = -12$.

$$4|x| = -12 \qquad \text{Rewrite the equation.}$$

$$|x| = -3 \qquad \text{Divide both sides by 4.}$$

Because absolute value is always nonnegative, no number has an absolute value of -3. So, this equation has *no solution.*

Without solving the equation, tell whether it has *one solution, two solutions,* or *no solution*. If the equation has one solution, tell whether the solution is *positive* or *negative*. Explain your reasoning.

1. $-5x = -16$

2. $\dfrac{n}{-5} = -12$

3. $\dfrac{g}{7} = -8$

4. $-12t = 100$

5. $-8|v| = -16$

6. $\dfrac{|k|}{-9} = 6$

7. $\left|\dfrac{x}{-5}\right| = -15$

8. $|-6p| = 42$

9. $|2.7u| = 10.8$

10. $\left|1\dfrac{1}{2}b\right| = -13\dfrac{4}{5}$

11. $3h = |8|$

12. $|-9|y = -12$

13. $\dfrac{|b|}{-3} = -7$

14. $\left|\dfrac{a}{-9}\right| = -2.5$

For each exercise number, use your answers and the key below to color the cell. Do not color the cells that have a zero in them.

One Positive Solution = Orange

One Negative Solution = Blue

Two Solutions = Green

No Solution = Green

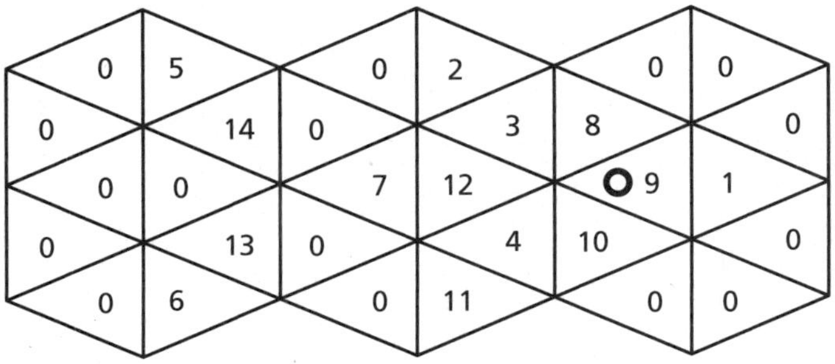

© Buki Toys LTD

Copyright © Big Ideas Learning, LLC
All rights reserved.

Name ______________________________ Date __________

13.4 Puzzle Time

Did You Hear About...

A	B	C	D	E	F
G	H	I	J	K	L
M					

Complete each exercise. Find the answer in the answer column. Write the word under the answer in the box containing the exercise letter.

−12 PLANET
−60 MOON
−14 GOOD
4 THE
−1.7 EARTH
−56 REALLY
9 ATMOSPHERE
$-13\frac{1}{2}$ THAT
$\frac{3}{11}$ ORBIT

Solve the equation.

A. $6x = 24$

B. $-7a = 35$

C. $-3g = -33$

D. $\frac{c}{4} = -8$

E. $\frac{z}{-12} = 5$

F. $\frac{2}{3}h = -9$

G. $-\frac{4}{5} = 2b$

H. $32 = -\frac{4}{7}y$

I. $-1.8m = 25.2$

J. $\frac{p}{3.7} = 5.1$

K. $20.3 = -2.9c$

L. $-12.6w = -16.38$

M. Tyler has \$11.25. How many ride tickets can he buy for himself and his friends if the ride tickets cost \$1.25 each?

8.3 EAT
11 ON
−5 RESTAURANT
−7 BUT
−32 THE
1.3 NO
18.87 FOOD
$-1\frac{3}{5}$ BAD
$-\frac{2}{5}$ HAS

Copyright © Big Ideas Learning, LLC
All rights reserved.

Activity 13.5 Start Thinking!
For use before Activity 13.5

In Sections 13.3 and 13.4, you learned how to solve one-step problems. Without using your notes, give an example and then describe to a partner how to solve each of the four types of one-step problems.

Activity 13.5 Warm Up
For use before Activity 13.5

Solve the equation. Check your solution.

1. $-9x = -108$
2. $x - 3 = -12$
3. $\frac{x}{5} = -8$
4. $12x = 144$
5. $x - 3.6 = 5.44$
6. $x + \frac{1}{2} = 4\frac{2}{3}$

Copyright © Big Ideas Learning, LLC
All rights reserved.

Lesson 13.5 Start Thinking!

For use before Lesson 13.5

Explain to a partner how to solve the equation $3x - 4 = 11$.

Lesson 13.5 Warm Up

For use before Lesson 13.5

Solve the equation. Check your solution.

1. $3x - 10 = 14$

2. $5x + 24 = -36$

3. $17 = 3x + 2$

4. $-9x + 17 = 71$

5. $4x + 13.2 = 2.6$

6. $-3x - 3.4 = 12.2$

Copyright © Big Ideas Learning, LLC
All rights reserved.

Name______________________________ Date__________

13.5 Practice A

Solve the equation. Check your solution.

1. $3k - 2 = 10$
2. $5p + 2 = -10$
3. $-4x + 3 = -11$
4. $12 = 2d + 3.2$
5. $-1 - 5h = 14$
6. $1.25r - 7 = 2.5$
7. $-4k + 3.6 = 7.8$
8. $6 + 2n = 3$
9. $4y - 16.3 = 53.1$
10. $\frac{1}{2}b + \frac{9}{4} = \frac{7}{4}$
11. $\frac{5}{6} + 3j = -\frac{2}{3}$
12. $-\frac{9}{10}p - 3 = \frac{3}{5}$

In Exercises 13–15, write an equation. Then solve.

13. It costs \$4 to enter the fair. Each ride costs \$2.50. You have \$21.50. How many rides can you go on?

14. The cable company charges a monthly fee of \$45. Each movie rental is \$1.99. You owe \$68.88. How many movies did you rent?

15. The perimeter of the rectangle is 24 feet. What is the width of the rectangle?

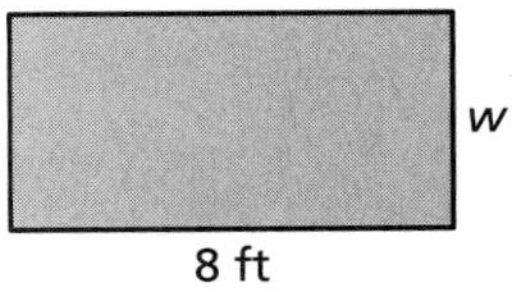

Solve the equation. Check your solution.

16. $7c - 2c = 45$
17. $3(k - 5) = -16$
18. $-2(m + 1) = 10$

19. The senior class has 412 students. They are assigned to different homerooms. There are 28 students in the smallest homeroom and the remaining 12 homerooms have the same number of students. How many students are in each of the remaining 12 homerooms?

20. You purchased paint for the rooms in your house. You have 1.5 cans of paint left. You painted 4 rooms and each room required 2 cans of paint. You spilled $\frac{1}{2}$ of a can of paint. How many cans of paint did you purchase?

    a. Solve the problem by working backwards.

    b. Solve the equation $\frac{x - 2}{4} = 2$. How does the answer compare to part (a)?

Copyright © Big Ideas Learning, LLC
All rights reserved.

Name ______________________________ Date __________

13.5 Practice B

Solve the equation. Check your solution.

1. $5k - 8 = 7$

2. $6b + 9 = -15$

3. $-3.2w - 2 = -4.5$

4. $13 - 2n = 27$

5. $25 = 4.5z + 12$

6. $5.25s - 2.01 = -8.94$

7. $84 = 51 - 14p$

8. $81 + 7t = 81$

9. $16 + 2.4c = 22.5$

10. $4h + \frac{1}{3} = \frac{3}{4}$

11. $\frac{1}{7}f - 5\frac{1}{2} = \frac{9}{14}$

12. $-\frac{1}{2}u + \frac{3}{5} = \frac{1}{6}$

In Exercises 13 and 14, write an equation. Then solve.

13. You purchased \$132.49 worth of wheels and bearings for your skateboards. The shop charges \$15 per board to install them. The total cost is \$192.49. How many skateboards will be repaired?

14. A music download service charges a flat fee each month and \$0.99 per download. The total cost for downloading 27 songs this month is \$42.72. How much is the flat fee?

Solve the equation. Check your solution.

15. $-5x - 2x + 3x = 9$

16. $-5(m + 4) = 27$

17. $-12(a - 2) = -50$

18. The perimeter of a triangle is 60 feet. One leg is 12 feet long. Of the two unknown sides, one of them is twice as long as the other. Find the lengths of the two unknown sides.

19. Sally picks seashells by the seashore. She lost 17 of them on her way home. She planned to fill 5 jars with the same amount of seashells in each. How many seashells did Sally pick?

a. You do not have enough information to solve this problem. The number of seashells in each jar is the same as the number portion of her street address, which is a 2-digit number. The first digit is 5. The last digit is 9 less than 3 times the first digit. How many seashells did Sally plan to put in each jar?

b. By working backwards, determine how many seashells Sally picked.

c. The 5 jars that Sally chose would not each hold that many seashells. In her search for a 6th jar, she discovered a few seashells in her pocket. What are possible values for the number of seashells in each of the 6 jars and the number of seashells discovered in her pocket, such that there are no seashells left over?

Copyright © Big Ideas Learning, LLC
All rights reserved.

Name____ Date__________

13.5 Enrichment and Extension

Solving Equations with Fractions

1. If you multiply each term by this number, the equation $\frac{3x}{2} - \frac{4}{5} = 5\frac{1}{5}$ will contain no fractions. What number could this be?

2. Are there other numbers you can multiply by to rewrite the equation in Exercise 1 without fractions? Explain.

3. What number do you think is best to use at the multiplier? Explain.

4. Why can you multiply each term and not change the solution of the equation?

5. Describe to someone how to rewrite an equation with fractions so that there are no fractions left in it.

6. Solve each equation by rewriting it without fractions first.

   a. $\frac{x}{8} - 5 = \frac{3}{4}$

   b. $\frac{x}{4} - \frac{1}{2} = 3\frac{1}{4}$

   c. $2\frac{1}{3} - \frac{x}{4} = \frac{5}{6}$

   d. $-\frac{2}{7} + \frac{x}{2} = \frac{9}{14}$

   e. $6\frac{7}{9} = \frac{2}{3} - 5x$

   f. $\frac{x + 10}{6} = \frac{2}{3}$

Copyright © Big Ideas Learning, LLC
All rights reserved.

Name ______________________________ Date __________

13.5 Puzzle Time

What Did One Bowling Ball Say To The Other Bowling Ball?

Write the letter of each answer in the box containing the exercise number.

Solve the equation.

1. $2c - 5 = 9$

2. $3m + 7 = -8$

3. $-7x - 3 = 12$

4. $15 = 4a + 3$

5. $5y - 6 = -20$

6. $9f + 3.6 = 10.8$

7. $-4p - 5.7 = 11.1$

8. $-20.3 = 6w + 3.1$

9. $2 + 5.3k = 18.43$

10. $7.8b - 2.14 = -42.7$

11. $\frac{1}{4}z - \frac{2}{7} = \frac{5}{7}$

12. $3 - \frac{r}{8} = -\frac{9}{2}$

13. $-\frac{1}{3} + 5e = -\frac{3}{4}$

14. $14d - 2d = -84$

15. $-5g - 13g = 54$

16. $-3(t - 8) = 32$

17. Kayla's age is 3 less than twice her brother's age. Kayla is 13 years old. How old is her brother?

18. Mario spent \$23.85 at the bookstore on one book and some magazines. The book cost \$12.60 and the magazines cost \$2.25 each. How many magazines did Mario buy?

19. Ethan planted a tree that is 37.5 inches tall. If the tree grows 3 inches each year, how long will it take for the tree to reach a height of 54 inches?

Answers

T. -5.2

N. 3

S. 5

M. 8

E. $-2\frac{4}{5}$

O. 3.1

L. $-2\frac{1}{7}$

T. 7

N. -7

O. $-\frac{1}{12}$

M. -5

I. 4

P. $-2\frac{2}{3}$

A. 0.8

O. -3.9

D. -3

L. 60

R. 5.5

O. -4.2

15	7	4	10		18	1	13	16		17	5		11	2		8	14		6		19	9	12	3

Copyright © Big Ideas Learning, LLC
All rights reserved.

Name______________________________ Date__________

Chapter 13

Technology Connection

For use after Section 13.5

Solving Multi-Step Equations

You may find it helpful to use a calculator to solve multi-step equations with large numbers or decimals.

EXAMPLE 1 Solve $19.39 + 4.8x = 34.51$.

SOLUTION

Step 1 Enter 34.51 [−] 19.39 to undo addition.

Step 2 Most calculators will store the answer, 15.12, so that you don't need to write it down. Just enter [÷] 4.8 to undo the multiplication.

ANSWER 3.15

You can use this method to solve equations that require more steps.

EXAMPLE 2 Solve $\frac{x}{1.2} + 1.8 = 9 - 1.45$.

SOLUTION

Step 1 Enter 9 [−] 1.45 to simplify the right side of the equation.

Step 2 Enter [−] 1.8 to undo addition.

Step 3 Enter [×] 1.2 to undo division.

ANSWER 6.9

Use a calculator to solve the equation.

1. $106y + 317 = 5299$
2. $0.93x - 0.904 = 0.2864$
3. $325.85 = 0.14m - 40.67$
4. $\frac{x}{52} - 73 = 36$
5. $\frac{w}{27} + 372 = 431$
6. $203 = \frac{m}{67} + 189$
7. $212x - 5216 = 9310 + 2646$
8. $4.838 - 1.202 = 1.85z - 3.209$
9. $41 + \frac{a}{109} = 22.3 + 21.5$
10. $0.65 + 9.85 = \frac{c}{0.65} + 2.6$

Copyright © Big Ideas Learning, LLC
All rights reserved.

Chapter 14

Copyright © Big Ideas Learning, LLC
All rights reserved.

Name______________________________ Date__________

Chapter 14 Ratios and Proportions

Dear Family,

An emergency evacuation plan is required in most commercial buildings. A good plan shows the locations of exits in the building as well as the locations of fire extinguishers and other emergency equipment. The plan is usually shown on a scale drawing of the building's floor plan.

Creating an emergency evacuation plan for your home is a good idea as well. You and your student can work together to make a scale drawing of your home.

Choose a scale that will make measurements relatively easy and allow the plan to fit on a single piece of paper. A common scale is $\frac{1}{4}$ inch for every foot. This scale will allow a building as large as 34 x 44 feet to fit on a letter-sized piece of paper. If your home won't fit within those dimensions, you can choose larger paper or a smaller scale—such as $\frac{1}{8}$ inch for every foot.

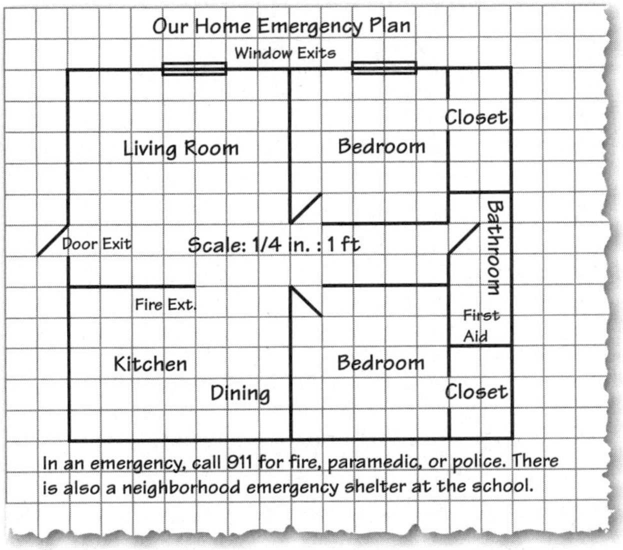

Mark the exits, fire extinguishers, and any alarms in red. If you have emergency medical equipment available, such as a first aid kit, mark those in blue.

Ask your student to help you with the following.

- Make measurements of each room in the home. Include measurements of doors and windows that will work as exits.
- Convert your measurements to the scale you have chosen.
- Draw the plan on $\frac{1}{4}$-inch graph paper. The sides of each square have a length of $\frac{1}{4}$-inch which makes it easier to use for $\frac{1}{4}$-inch and $\frac{1}{8}$-inch scales.

It's a good idea to include other information on the plan as well, such as the numbers for fire, paramedic, and police services (provided by 911 in many towns). Work with your student to decide what information should be included.

Being prepared will give you great peace of mind!

Copyright © Big Ideas Learning, LLC
All rights reserved.

Nombre ______________________________ Fecha ________

Capítulo 14 Razones y Proporciones

Estimada Familia:

En la mayoría de edificios comerciales se requiere un plano de evacuación de emergencia. Un buen plano muestra las ubicaciones de las salidas del edificio, así como las ubicaciones de los extintores de incendios y otros equipos de emergencia. El plano generalmente se muestra en un dibujo a escala del plano del piso del edificio.

Crear un plano de evacuación de emergencia para su hogar también puede ser una buena idea. Usted y su estudiante pueden trabajar juntos para hacer un dibujo a escala de su hogar.

Elijan una escala en la que puedan hacerse medidas de manera relativamente fácil y permita que el plano quepa en una sola hoja de papel. Una escala común es $\frac{1}{4}$ pulgada por cada pie. Esta escala permite que un edificio tan grande como de 34 x 44 pies quepa dentro de una hoja de papel tamaño carta. Si su hogar no cupiera dentro de esas dimensiones, puede elegir un papel más grande o una escala más pequeña—como por ejemplo $\frac{1}{8}$ pulgada por cada pie.

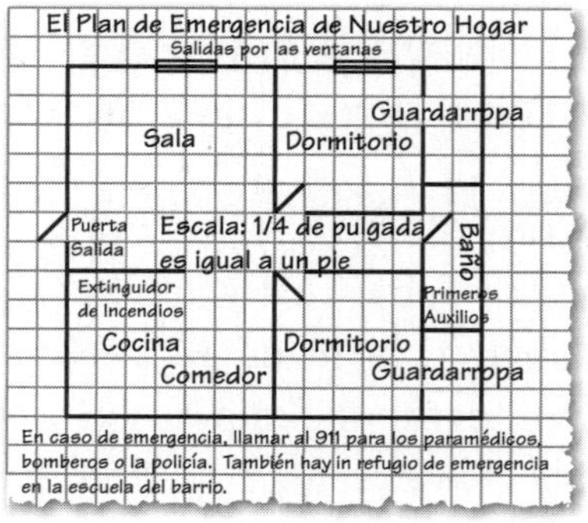

Marque las salidas, extintores y alarmas en rojo. Si tiene disponible equipo médico de emergencia, como por ejemplo un equipo de primeros auxilios, márquelo en azul.

Pida a su estudiante que lo ayude con lo siguiente:

- Hagan medidas de cada habitación de la casa. Incluyan medidas de las puertas y ventanas, ya que funcionarán como salidas.
- Conviertan sus medidas en la escala elegida.
- Dibujen el plano en un papel para gráficos de $\frac{1}{4}$ pulgada. Los lados de cada cuadrado miden $\frac{1}{4}$ pulgada de largo, lo que los hace más fácil de usar con las escalas de $\frac{1}{4}$ pulgada y $\frac{1}{8}$ pulgada.

Es una buena idea incluir también otra información en el plano, como los teléfonos de los servicios de bomberos, paramédicos y policías (proporcionados por 911 en muchas ciudades). Trabaje con su estudiante para decidir qué información debe incluirse.

¡Estar preparado le dará una gran tranquilidad!

Copyright © Big Ideas Learning, LLC
All rights reserved.

Activity 14.1

Start Thinking!

For use before Activity 14.1

You are planning a trip to Atlanta, Georgia. Discuss with a partner the factors you need to consider to determine if it would be better to fly or to drive. Determine which method of transportation you would use and explain why.

Activity 14.1

Warm Up

For use before Activity 14.1

Convert the measurement.

1. 30 min = __?__ h

2. 4 h = __?__ min

3. 15 sec = __?__ min

4. 60 h = __?__ days

5. 3 days = __?__ h

6. 1 wk = __?__ h

Copyright © Big Ideas Learning, LLC
All rights reserved.

Lesson 14.1 Start Thinking!

For use before Lesson 14.1

You and your sister go to a store. You buy 3 erasers for $1.00. Your sister buys 4 erasers for $1.25. Explain who gets the better deal.

Lesson 14.1 Warm Up

For use before Lesson 14.1

Find the product. List the units.

1. $6 \text{ h} \times \frac{\$7}{\text{h}}$

2. $2 \text{ gal} \times \frac{\$24}{\text{gal}}$

3. $8 \text{ h} \times \frac{25 \text{ mi}}{\text{h}}$

4. $9 \text{ mo} \times \frac{\$650}{\text{mo}}$

5. $12 \text{ lb} \times \frac{\$2.50}{\text{lb}}$

6. $6 \text{ yr} \times \frac{35 \text{ in.}}{\text{yr}}$

Copyright © Big Ideas Learning, LLC
All rights reserved.

Name____ Date__________

14.1 Practice A

Find the product. List the units.

1. $12 \text{ h} \times \frac{\$5}{\text{h}}$ **2.** $6 \text{ oz} \times \frac{\$0.59}{\text{oz}}$ **3.** $9 \text{ h} \times \frac{70 \text{ mi}}{\text{h}}$

Write the ratio as a fraction in simplest form.

4. 12 to 15 **5.** 24 : 9 **6.** 14 tetras : 6 angelfish

Find the unit rate.

7. 360 miles in 6 hours **8.** 18 bowlers on 6 lanes **9.** $28 for 7 people

Use the ratio table to find the unit rate with respect to the specified units.

10. Laps per minute

Minutes	0	2	4	6
Laps	0	1	2	3

11. Grams of protein per serving

Servings	0	1	2	3
Grams of Protein	0	15	30	45

12. At 9 A.M. you have run 2 miles. At 9:24 A.M. you have run 5 miles. What is your running rate in minutes per mile?

13. Are the two statements equivalent? Explain your reasoning.

- The ratio of orange to blue is 3 to 4.
- The ratio of blue to orange is 12 to 9.

14. There are 234 students in 9 different classrooms. What is the ratio of students to classrooms?

15. Dishwasher detergent is sold in individual packs. It is sold in 20-, 60-, and 90-pack containers.

a. Which container do you think has the lowest unit rate of dollars per pack? Why?

b. The 20-pack container sells for $5.49. What is the unit rate in dollars per pack? Round your answer to the nearest cent.

c. The 60-pack container sells for $10.97. What is the unit rate in dollars per pack? Round your answer to the nearest cent.

d. The 90-pack container sells for $18.95. What is the unit rate in dollars per pack? Round your answer to the nearest cent.

e. Which container has the lowest unit rate? How does this compare with your answer in part (a)?

Copyright © Big Ideas Learning, LLC
All rights reserved.

Name ______________________________ Date __________

14.1 Practice B

Write the ratio as a fraction in simplest form.

1. 35 to 63

2. 10.8 seconds : 36 feet

3. 198 women to 110 men

4. 1000 songs : 2 megabytes

5. 26.1 miles : 3.6 hours

6. 12 completions to 28 attempts

Find the unit rate.

7. \$5.40 for 24 cans

8. \$1.29 for 20 ounces

9. 50 meters in 27.5 seconds

10. There are 16 bacteria in a beaker. Four hours later there are 228 bacteria in the beaker. What is the rate of change per hour in the number of bacteria?

11. The table shows nutritional information for three energy bars.

a. Which has the most protein per calorie?

b. Which has the least sugar per calorie?

c. Which has the highest rate of sugar to fiber?

Energy Bar	Calories	Protein	Fiber	Sugar
A	220	20 g	12 g	14 g
B	130	12 g	8 g	10 g
C	140	4 g	9 g	9 g

d. Compare bar A with bar B. Which nutritional item do you think has the highest ratio: calories, protein, fiber, or sugar?

e. Calculate the ratios in part (d). Which one has the highest ratio?

12. The graph shows the cost of buying scoops of gelato.

a. What does the point $(4, 6)$ represent?

b. What is the unit cost?

c. What is the cost of 12 scoops?

d. Explain how the graph would change if the unit rate was \$1.75 per scoop.

e. How would the coordinates of the point in part (a) change if the unit rate was \$1.75 per scoop?

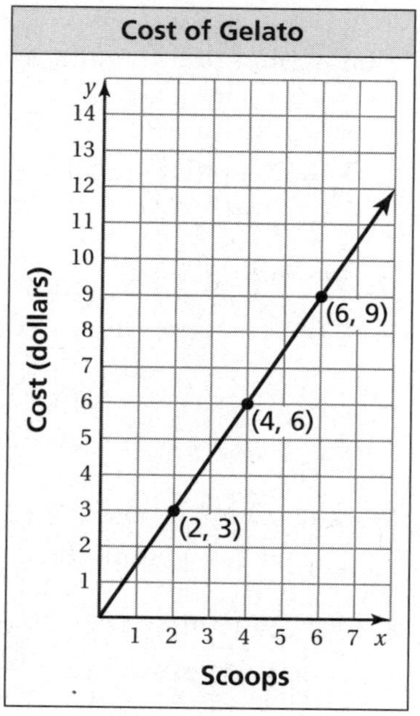

Write a situation for the ratio.

13. $\frac{9}{5}$

14. 2 : 3

Copyright © Big Ideas Learning, LLC
All rights reserved.

Name____ Date__________

14.1 Enrichment and Extension

Ratios, Rates, and On-the-Job Decisions

Sally Smith has been offered two employment opportunities. Help her decide which job is better by answering the following questions.

1. Job A will pay \$32,448 per year. How much money would Sally be making per hour? Assume pay is based on fifty-two 40-hour weeks.

2. Job B pays \$14.80 per hour. How much money would she make in a year? Assume the pay is based on fifty-two 40-hour weeks.

3. Which job pays better? Explain your reasoning.

4. Sally lives 18 miles from Job A. A work week is five days. How many miles would she have to drive each week just to get to and from work?

5. For Job A, Sally would have to put 7.5 gallons of gas in her tank every 3 days that she drives to and from work only. How many gallons of gas would she use in a five-day work week?

6. Based on your answers to Exercises 4 and 5, what is Sally's gas mileage? (*Hint:* Gas mileage is a unit rate that is calculated as miles per gallon.)

7. If gas costs \$4 per gallon, how much will gas cost Sally per day just to get to and from Job A per day? per week? per year?

8. The ratio of the length of Sally's drive to Job A to the length of Sally's drive to Job B is 2 : 3. How much money would she save on gas in a week with Job A as opposed to Job B?

9. Other than distance and cost of gas, what other factors should Sally consider when comparing commutes to work?

10. Macaroni and cheese is one of Sally's favorite lunch foods. In the cafeteria at Job A, $1\frac{1}{4}$ cups of macaroni and cheese contains 6.5 grams of fat. The macaroni and cheese at Job B contains 10 grams of fat per pint. Which one has a lower fat content? Explain your reasoning.

11. Both places sell macaroni and cheese by weight. Job A's cafeteria charges \$7.25 per pound and Job B's cafeteria charges \$0.58 per ounce. Which cafeteria has the cheaper macaroni and cheese? Explain your reasoning.

12. Based on the information on this page, would you recommend that Sally take Job A or Job B? Explain your reasoning.

13. Other than pay, transportation costs, and cafeteria selections, what other factors should Sally consider when choosing a job?

Copyright © Big Ideas Learning, LLC
All rights reserved.

Name ______________________________ Date __________

14.1 Puzzle Time

What Do You Get If You Cross A Duck With A Firework?

Write the letter of each answer in the box containing the exercise number.

Find the product.

1. $4 \text{ tbsp} \times \frac{20 \text{ cal}}{\text{tbsp}}$

P. 5 cal **Q.** 80 cal **R.** 5 tbsp

2. $3 \text{ lb} \times \frac{\$1.29}{\text{lb}}$

E. \$0.43 **F.** \$3.87 **G.** 3.87 lb

3. $4 \text{ gal} \times \frac{17.5 \text{ mi}}{\text{gal}}$

C. 70 mi **D.** 35 mi **E.** 70 gal

4. $40 \text{ h} \times \frac{\$8.50}{\text{h}}$

T. 340 h **U.** \$340 **V.** \$300

Write the ratio as a fraction.

5. 12 to 36

I. $\frac{1}{6}$ **J.** $\frac{1}{4}$ **K.** $\frac{1}{3}$

6. 15 : 75

E. $\frac{1}{5}$ **F.** $\frac{1}{4}$ **G.** $\frac{1}{3}$

7. 10 out of 25

E. $\frac{2}{5}$ **F.** $\frac{1}{5}$ **G.** $\frac{1}{25}$

8. 52 males to 28 females

P. $\frac{27}{14}$ **Q.** $\frac{7}{13}$ **R.** $\frac{13}{7}$

Find the unit rate.

9. 48 cups in 12 quarts

R. $\frac{4 \text{ c}}{\text{qt}}$ **S.** $\frac{3 \text{ c}}{\text{qt}}$ **T.** $\frac{2 \text{ c}}{\text{qt}}$

10. \$17.85 for 3 pounds

A. $\frac{\$5.95}{\text{lb}}$ **B.** $\frac{\$5.56}{\text{lb}}$ **C.** $\frac{\$3.65}{\text{lb}}$

11. 26.2 miles in 4 hours

A. $\frac{6.55 \text{ mi}}{\text{h}}$ **B.** $\frac{7.5 \text{ mi}}{\text{h}}$ **C.** $\frac{6.5 \text{ mi}}{\text{h}}$

12. \$12.60 for 3 boxes

G. $\frac{\$3.15}{\text{box}}$ **H.** $\frac{\$4.15}{\text{box}}$ **I.** $\frac{\$4.20}{\text{box}}$

10		2	12	8	6		1	4	11	3	5	7	9
						-							

Copyright © Big Ideas Learning, LLC
All rights reserved.

Activity 14.2 Start Thinking!
For use before Activity 14.2

Give an example of two fractions that are equivalent. Show that they are equivalent. Give an example of two fractions that are not equivalent. Explain why they are not equivalent.

Activity 14.2 Warm Up
For use before Activity 14.2

Write the fraction in simplest form. Do not change improper fractions to mixed numbers.

1. $\frac{24}{48}$

2. $\frac{64}{48}$

3. $\frac{30}{24}$

4. $\frac{9}{21}$

5. $\frac{45}{63}$

6. $\frac{10}{41}$

7. $\frac{144}{12}$

8. $\frac{50}{35}$

Copyright © Big Ideas Learning, LLC
All rights reserved.

Lesson 14.2 Start Thinking!
For use before Lesson 14.2

You go to the grocery store to buy cereal. How do you determine which cereal brand is the better buy?

Lesson 14.2 Warm Up
For use before Lesson 14.2

Tell whether the two rates form a proportion.

1. 5 feet in 4 hours; 15 feet in 12 hours
2. 8 pages in 40 minutes; 15 pages in 70 minutes
3. 3 pounds for \$3.75; 5 pounds for \$6.50
4. 2 cups in 4 servings; 5 cups in 10 servings

Copyright © Big Ideas Learning, LLC
All rights reserved.

Name____________________________________ Date__________

14.2 Practice A

Tell whether the ratios form a proportion.

1. $\frac{1}{4}, \frac{3}{12}$

2. $\frac{1}{7}, \frac{4}{28}$

3. $\frac{2}{5}, \frac{30}{80}$

4. $\frac{18}{24}, \frac{15}{20}$

5. $\frac{35}{16}, \frac{5}{2}$

6. $\frac{5}{7}, \frac{35}{49}$

7. $\frac{15}{21}, \frac{40}{56}$

8. $\frac{33}{63}, \frac{26}{42}$

9. $\frac{54}{10}, \frac{81}{15}$

Tell whether the two rates form a proportion.

10. 8 feet in 15 seconds; 16 feet in 40 seconds

11. 28 people in 4 rooms; 63 people in 9 rooms

12. 14 girls to 6 boys; 35 girls to 15 boys

13. 45 marbles in 9 bags; 150 marbles in 36 bags

14. You can run 4 laps in 10 minutes. Your friend can run 6 laps in 15 minutes. Are these rates proportional? Explain.

Tell whether the ratios form a proportion.

15. $\frac{7}{4}, \frac{17.5}{10}$

16. $\frac{1.5}{6}, \frac{2}{8}$

17. $\frac{8}{5}, \frac{68}{45}$

18. You get \$27 to spend at the mall for doing 6 chores. Your friend gets \$36 for doing 8 chores.

a. What is your pay rate?

b. What is your friend's pay rate?

c. Are the pay rates equivalent? Explain.

19. You can buy 4 tickets for \$75 or 5 tickets for \$94. Are the costs proportional? If not, rewrite one of the rates so the costs are proportional.

20. A recipe requires a ratio of 4 potatoes to 6 carrots. You accidentally use 5 potatoes with 6 carrots. What is the least number of potatoes and carrots that you can add to get the correct ratio of potatoes to carrots?

Copyright © Big Ideas Learning, LLC
All rights reserved.

Name ______________________________ Date __________

14.2 Practice B

Tell whether the ratios form a proportion.

1. $\frac{25}{16}, \frac{65}{56}$

2. $\frac{30}{75}, \frac{24}{60}$

3. $\frac{27}{48}, \frac{108}{192}$

Tell whether the two rates form a proportion.

4. \$24 for 16 burgers; \$15 for 10 burgers

5. 10 used books for \$4.50; 15 used books for \$6.00

6. 125 horsepower motor for an 18-foot boat; 225 horsepower motor for a 32-foot boat

Tell whether the ratios form a proportion.

7. $\frac{28.5}{42}, \frac{19}{28}$

8. $\frac{3.5}{4}, \frac{11.9}{13.6}$

9. $\frac{124}{98}, \frac{315}{225}$

10. The seventh-grade band has 15 drummers and 12 trumpet players. The eighth-grade band has 10 drummers and 8 trumpet players. Do the ratios form a proportion? Explain.

11. One mixture contains 6 fluid ounces of water and 10 fluid ounces of vinegar. A second mixture contains 9 fluid ounces of water and 12 fluid ounces of vinegar. Are the mixtures proportional? If not, how much water or vinegar would you add to the second mixture so that they are proportional?

12. A wholesale warehouse buys pairs of sandals to sell.

a. The warehouse can purchase 5 pairs of sandals for \$65. What is the cost rate?

b. The warehouse can purchase 8 pairs of sandals for \$96. What is the cost rate?

c. The warehouse can purchase 10 pairs of sandals for \$126.50 and will get one free pair. What is the cost rate?

d. Are any of the cost rates proportional? Explain.

e. Your buyer is to purchase 40 pairs of sandals. Use any combination of parts (a), (b), and (c) for your buyer to purchase the 40 pairs of sandals at the lowest possible cost.

Find the value of *x* so that the ratios form a proportion.

13. $\frac{3}{7}, \frac{x}{21}$

14. $\frac{16}{12}, \frac{20}{x}$

Copyright © Big Ideas Learning, LLC
All rights reserved.

Name____ Date__________

14.2 Enrichment and Extension

The Big Paint "Mix Up"

Oh no! The paint mixing machine has gone crazy! Can you help figure out how to fix these batches by adding the least amount of paint possible? Otherwise all the paint will go to waste. For each situation, tell how much of one or more colors should be added to each batch. The machine can only measure in cups, pints, quarts, and gallons.

1. *Brick Red* is supposed to be 7 parts red to 2 parts blue. Today, the machine mixed 35 quarts of red with 5 quarts of blue.

2. *Ocean Blue* is supposed to be 8 parts blue to 3 parts yellow. Today, the machine mixed 55 quarts of blue with 15 quarts of yellow.

3. *Sour Apple* is supposed to be 9 parts yellow to 2 parts blue. Today, the machine mixed 2 gallons of yellow with 2 quarts of blue.

4. *Midnight Navy* is supposed to be 7 parts blue to 3 parts red. Today, the machine mixed 3 quarts of blue with 1 pint of red.

5. *Sunset Yellow* is supposed to be 6 parts yellow to 1 part red. Today, the machine mixed 70 pints of yellow with 11 pints of red.

6. The last correct batch of *Burnt Orange* had 15 quarts of yellow and 30 quarts of red. Today, the machine mixed 42 gallons of yellow with 72 gallons of red.

7. The last correct batch of *Perfectly Grape* had 20 pints of red and 24 pints of blue. Today, the machine mixed 28 quarts of red with 32 quarts of blue.

8. The last correct batch of *Mustard Brown* had 96 pints of yellow, 72 pints of red, and 24 pints of blue. Today the machine mixed 45 cups of yellow, 20 cups of red, and 8 cups of blue.

Copyright © Big Ideas Learning, LLC
All rights reserved.

Name ______________________________ Date __________

14.2 Puzzle Time

What Can You Hold Without Ever Touching?

For each exercise, circle the letter in the columns under Yes or No to indicate the correct answer. The circled letters will spell the answer to the riddle.

	Yes	No
1.	Y	T
2.	A	O
3.	U	H
4.	R	S
5.	C	B
6.	O	R
7.	E	O
8.	A	L
9.	T	M
10.	S	H

Tell whether the ratios form a proportion.

1. $\frac{2}{5}, \frac{8}{20}$

2. $\frac{3}{7}, \frac{6}{13}$

3. $\frac{5}{6}, \frac{15}{18}$

4. $\frac{18}{24}, \frac{12}{16}$

Tell whether the two rates form a proportion.

5. 55 miles in 1 hour; 450 miles in 8 hours

6. \$3.00 for 32 ounces of strawberries; \$1.75 for 24 ounces of strawberries

7. 45 baskets in 85 shots; 54 baskets in 102 shots

8. 18 push-ups in 60 seconds; 27 push-ups in 90 seconds

9. One type of cereal has 2 grams of protein per 1-cup serving. Another cereal has 1 gram of protein per half-cup serving. Do these rates form a proportion?

10. A 50-fluid ounce bottle of laundry detergent washes 32 loads of laundry. A 100-fluid ounce bottle washes 60 loads of laundry. Are they proportional? Do these rates form a proportion?

Copyright © Big Ideas Learning, LLC
All rights reserved.

Extension 14.2 Start Thinking!

For use before Extension 14.2

Your friend spent \$7 in 2 days on lunch. He says that he will need \$21 for 6 days to spend on lunch. Is your friend correct? Explain your reasoning.

Extension 14.2 Warm Up

For use before Extension 14.2

Tell whether *x* and *y* are proportional.

1.

x	1	2	3	4
y	4	8	12	16

2.

x	2	3	4	5
y	2	4	6	8

Copyright © Big Ideas Learning, LLC
All rights reserved.

Name ______________________ Date __________

Extension 14.2 Practice

Interpret each plotted point in the graph of the proportional relationship.

1.

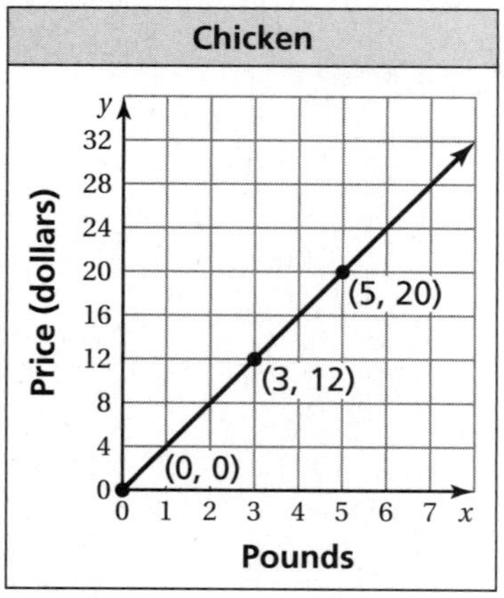

2.

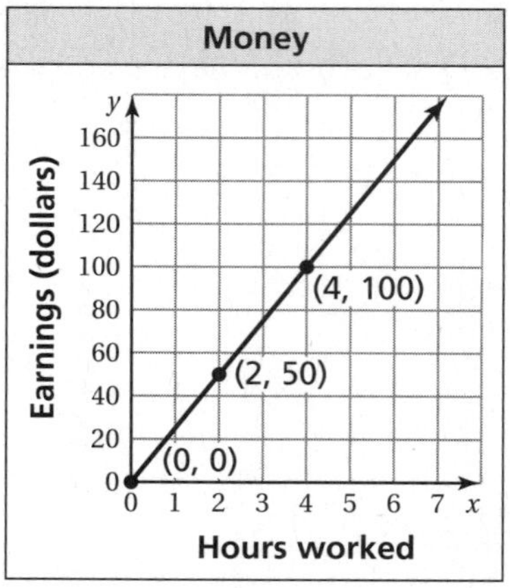

The graph of a proportional relationship passes through the given points. Find *y*.

3. $(4, 8), (1, y)$

4. $(3, 21), (1, y)$

5. $(1.5, 9), (1, y)$

6. $(3.5, 14), (1, y)$

7. Two classes have car washes to raise money for class trips. A portion of the earnings will pay for using the two locations for the car washes. The graph shows that the trip earnings of the two classes are proportional to the car wash earnings.

a. Express the trip earnings rate for each class as a percent.

b. What trip earnings does Class A receive for earning \$75 from the car wash?

c. How much less does Class B receive than Class A for earning \$75 from the car wash?

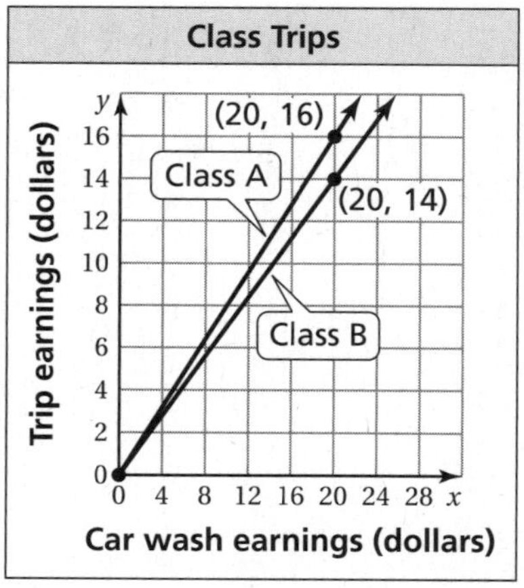

Copyright © Big Ideas Learning, LLC
All rights reserved.

Activity 14.3 Start Thinking!

For use before Activity 14.3

How do you determine if two ratios are proportional? Explain giving an example of ratios that are proportional.

Activity 14.3 Warm Up

For use before Activity 14.3

Write two equivalent ratios for the ratio given.

1. $\frac{10}{15}$

2. $\frac{11}{20}$

3. $\frac{6}{14}$

4. $\frac{4}{10}$

5. $\frac{12}{20}$

6. $\frac{6}{15}$

7. $\frac{8}{9}$

8. $\frac{14}{24}$

Copyright © Big Ideas Learning, LLC
All rights reserved.

Lesson 14.3 Start Thinking!
For use before Lesson 14.3

You scored 15 points in 3 minutes. Your cousin scored 25 points in 5 minutes.

Are the scores proportional? Explain.

Lesson 14.3 Warm Up
For use before Lesson 14.3

Write a proportion to find how many points a student needs to have on the test to get the given score.

1. test worth 100 points; test score of 85%
2. test worth 50 points; test score of 74%
3. test worth 25 points; test score of 80%
4. test worth 110 points; test score of 90%

Copyright © Big Ideas Learning, LLC
All rights reserved.

Name____ Date__________

14.3 Practice A

Write a proportion to find how many points a student needs to earn on the test to get the given score.

1. test worth 70 points; test score of 90%
2. test worth 30 points; test score of 72%

Write a proportion to find how many free throws a player needs to get the given score.

3. 15 free-throw attempts; free-throw score of 60%

4. 24 free-throw attempts; free-throw score of 75%

Use the table to write a proportion.

5.

	August	September
Hurricanes	2	1
Storms	6	n

6.

	Day 1	Day 2
Wins	w	8
Races	21	12

7. The county requires 2 teachers for every 45 students. Write a proportion that gives the number t of teachers needed for 315 students.

Solve the proportion.

8. $\frac{2}{3} = \frac{a}{15}$

9. $\frac{4}{7} = \frac{44}{m}$

10. $\frac{d}{6} = \frac{72}{48}$

11. A paint color requires the ratio of green paint to yellow paint to be 4 : 9.

    a. A container of this paint has 36 pints of yellow paint. Write a proportion that gives the number g of pints of green paint in the container.

    b. How many pints of green paint are in the container?

    c. How many *gallons* of paint are in the container altogether?

12. An orchestra has 10 cellists.

    a. There are 3 violin players for every cellist in the orchestra. How many violin players are there?

    b. There are 6 viola players for every 5 cellists in the orchestra. How many viola players are there?

    c. What is the ratio of viola players to violin players? Give your answer in simplest form.

13. Give two possible pairs of values for p and q: $\frac{2}{5} = \frac{p}{q}$.

Copyright © Big Ideas Learning, LLC
All rights reserved.

Name ______________________________ Date __________

14.3 Practice B

In Exercises 1 and 2, write a proportion to find how many strikes a bowler needs to get the given score.

1. 32 strike attempts; strike score of 75%

2. 80 strike attempts; strike score of 95%

3. Describe and correct the error in writing the proportion.

X

	Day 1	Day 2
Length	3.1	15.5
Height	h	45

$$\frac{15.5}{h} = \frac{3.1}{45}$$

4. There are 3 referees for every 16 players. Write a proportion that gives the number of referees r for 128 players.

Solve the proportion.

5. $\frac{5}{12} = \frac{x}{36}$

6. $\frac{20}{3.4} = \frac{800}{y}$

7. $\frac{2.8}{r} = \frac{70}{3}$

8. $\frac{48}{15} = \frac{k}{37.5}$

9. $\frac{21}{p} = \frac{252}{78}$

10. $\frac{2.3}{1.6} = \frac{46}{w}$

11. A recipe calls for $\frac{3}{4}$ cup of sugar and $\frac{1}{2}$ cup of brown sugar. You are reducing the recipe. You will use $\frac{1}{6}$ cup of brown sugar. How much sugar will you use?

12. A calculator has 50 keys in five colors: gray, black, blue, yellow, and green.

    a. There are 6 gray keys for every 7 blue keys. Write the possible ratios for gray to blue keys.

    b. There are 6 gray keys for every 11 black keys. Write the possible ratios for gray to black keys.

    c. There are 6 gray keys for every 11 black keys. Also, the number of black keys is 2 less than twice the number of gray keys. Use your answer to part (b) to determine how many gray keys and how many black keys there are.

    d. There is 1 yellow key for every 1 green key. How many keys of each color are there?

Copyright © Big Ideas Learning, LLC
All rights reserved.

Name______________________________ Date__________

14.3 Enrichment and Extension

Using Slope to Design Handicap Ramps

According to the Americans with Disabilities Act, the maximum rise to run ratio for all handicap ramps is 1 : 12. Also, there must be a landing, or flat rest area, at the top and bottom of all ramps. These landings must be at least 60 inches in length. If there is a change in direction on the ramp, there must be a flat landing that is at least 60 inches by 60 inches.

1. A ramp with a rise of 5 inches must have a run of at least how many inches? how many feet?

2. Why do you think there is a maximum slope for handicap ramps?

3. What would be the advantages and disadvantages of making a ramp that is less steep?

4. Your school is being renovated and a new handicap ramp must be built. The bottom of the new front door is 20 inches above the sidewalk. What is the minimum length (in feet) of a straight ramp to this door, including the landings at the top and bottom?

5. An existing ramp on the old part of the school has a rise of 10.5 inches and a run of 12 feet 3 inches, not including the landings. Does it meet the requirements or will it have to be changed? Explain.

6. At the main bus entrance, there is a ramp with a change in direction. It has a sloping part with a rise of 5 inches and a run of 6 feet 8 inches, a landing where there is a 90 degree turn, and another sloping part with a rise of 7.5 inches and a run of 13 feet 9 inches. Which part of the ramp is steeper? Why might the ramp have been designed in this way?

7. Another ramp with an overall rise of 30 inches has to be built to replace the ramp described in Exercise 6. It must have two sloping parts that should have the same slopes as the two sloping parts of the ramp in Exercise 6. Both parts should have a rise of 15 inches, with a landing in between. Find the lengths of both sloping parts of the ramp.

8. Find a place in your school, home, neighborhood, or elsewhere that could use a ramp. Measure the rise of that area. Then design a ramp, and draw a picture or describe it in detail. Be sure to tell how long each portion of the ramp should be as well as the overall length.

Copyright © Big Ideas Learning, LLC
All rights reserved.

Name ______________________________ Date __________

14.3 Puzzle Time

Who Do Whales Go To See When Their Teeth Need To Be Fixed?

Write the letter of each answer in the box containing the exercise number.

Solve the proportion.

1. $\frac{3}{5} = \frac{h}{20}$

2. $\frac{4}{7} = \frac{24}{a}$

3. $\frac{1}{x} = \frac{5}{45}$

4. $\frac{2}{13} = \frac{m}{39}$

5. $\frac{t}{28} = \frac{3}{4}$

6. $\frac{18}{21} = \frac{6}{w}$

7. $\frac{3.4}{4.2} = \frac{r}{21}$

8. $\frac{1.5}{2.5} = \frac{6}{s}$

9. $\frac{q}{1.7} = \frac{16}{17}$

10. $\frac{2.2}{n} = \frac{44}{66}$

11. You need 3 tickets for one go-kart ride. How many tickets do you need for five go-kart rides?

12. Yesterday you downloaded 3 songs for $2.97. How many songs did you download today for $3.96?

13. There are 32 students in the school play. The ratio of girls to all students in the play is 5 : 8. How many girls are in the play?

14. Two out of three vehicles in a parking lot are SUVs. There are 18 SUVs in the parking lot. How many vehicles are in the parking lot?

Answers

I.	27	**T.**	10
O.	12	**H.**	9
T.	1.6	**N.**	17
D.	4	**C.**	20
E.	7	**T.**	3.3
A.	42	**R.**	21
S.	6	**O.**	15

9	3	6		11	5	13	2		12	1	7	10	14	4	8
								-							

Copyright © Big Ideas Learning, LLC
All rights reserved.

Activity 14.4 Start Thinking!
For use before Activity 14.4

Give a real-world example of things that are proportional.

Activity 14.4 Warm Up
For use before Activity 14.4

Find the unit rate.

1. 140 students in 4 buses

2. 605 miles in 5 hours

3. 312 pages in 4 hours

4. 54 seats in 2 rows

5. 1023 miles in 3 hours

6. 544 pages in 4 days

Copyright © Big Ideas Learning, LLC
All rights reserved.

Lesson 14.4

Start Thinking!

For use before Lesson 14.4

Describe how someone who works in a bakery might use proportions.

Lesson 14.4

Warm Up

For use before Lesson 14.4

Use the Cross Products Property to solve the proportion.

1. $\frac{a}{4} = \frac{15}{12}$

2. $\frac{11}{4} = \frac{m}{16}$

3. $\frac{2}{3} = \frac{v}{18}$

4. $\frac{8}{n} = \frac{11}{22}$

5. $\frac{t}{10} = \frac{15}{25}$

6. $\frac{10}{3} = \frac{8}{k}$

Copyright © Big Ideas Learning, LLC
All rights reserved.

Name______________________________ Date__________

14.4 Practice A

Use multiplication to solve the proportion.

1. $\frac{7}{4} = \frac{y}{28}$

2. $\frac{d}{48} = \frac{3}{4}$

3. $\frac{j}{8} = \frac{35}{56}$

Use the Cross Products Property to solve the proportion.

4. $\frac{14}{21} = \frac{b}{9}$

5. $\frac{10}{p} = \frac{6}{9}$

6. $\frac{55}{4} = \frac{h}{6}$

7. Eighteen oranges are packaged in 3 containers. How many oranges are packaged in 7 containers?

8. It costs $270 for 3 people to go on a fishing trip. How much does it cost for 10 people to go on the fishing trip?

Solve the proportion.

9. $\frac{3x}{10} = \frac{9}{4}$

10. $\frac{5x}{3} = \frac{80}{12}$

11. $\frac{7}{2} = \frac{x + 1}{6}$

12. Tell whether the statement is *true* or *false*. Explain.

$$\text{If } \frac{p}{q} = \frac{3}{5}, \text{ then } \frac{5}{p} = \frac{3}{q}.$$

13. The dimensions of a miniature model are proportional to the dimensions of the actual building.

    a. A wall that is 12 feet high on the building is 36 centimeters high on the model. Find the height on the model of a door that is 9 feet high on the building.

    b. Use a different method than the one you used in part (a) to find the number of centimeters on the model for a window that is 3 feet wide.

14. The ratio of men to women at a lecture is 2 to 5. A total of 63 people are at the lecture. How many are men? Explain how you found your answer.

15. The distance traveled (in feet) is proportional to the number of seconds. Find the values of x, y, and z.

Feet	3	x	15	z
Seconds	5	65	y	3.5

16. You train for a race by running at a speed of 6 miles per hour.

    a. At this speed, how many *minutes* does it take you to run 3.2 miles?

    b. On race day, you run 3.2 miles in 30 minutes. What is your speed in miles per hour?

Copyright © Big Ideas Learning, LLC
All rights reserved.

Name ______________________________ Date __________

14.4 Practice B

Use either multiplication or the Cross Products Property to solve the proportion.

1. $\frac{16}{9} = \frac{q}{36}$ **2.** $\frac{57}{r} = \frac{38}{18}$ **3.** $\frac{50}{14} = \frac{w}{98}$

4. $\frac{n}{121} = \frac{52}{22}$ **5.** $\frac{96}{45} = \frac{b}{20}$ **6.** $\frac{24}{f} = \frac{15}{36}$

7. Three shirts cost \$9.99. How much does it cost for 8 shirts?

Solve the proportion.

8. $\frac{8x}{13} = \frac{64}{52}$ **9.** $\frac{c - 3}{6} = \frac{7}{3}$ **10.** $\frac{20}{9} = \frac{10}{s + 2}$

11. The number of grams of protein is proportional to the number of servings.

a. How many servings provide 32.5 grams of protein?

b. Use a different method than part (a) to find how many servings provide 52 grams of protein.

c. How many grams of protein will 7 servings provide?

d. If 1 serving is equal to $\frac{3}{4}$ cup, how many cups does it take to get 19.5 grams of protein?

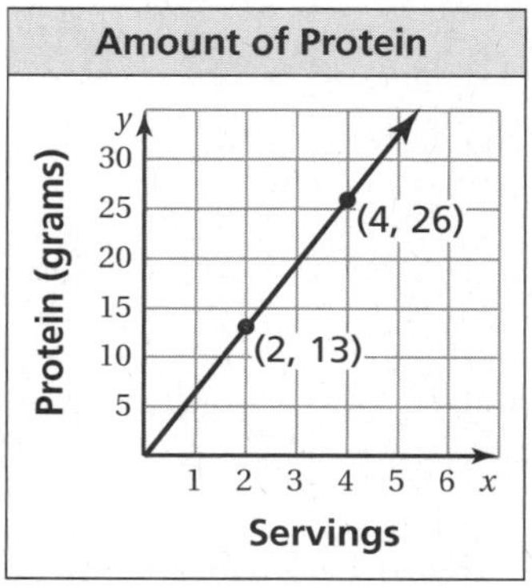

12. One day 176 people visited a small art museum. The ratio of members to nonmembers that day was 5 to 11. How many people who visited the museum that day were nonmembers?

13. Solve the proportion: $\frac{|h|}{3} = \frac{4}{9}$.

14. One gallon of water weighs about 8.34 pounds.

a. How much does 3.5 gallons of water weigh?

b. One inch of rain on a square foot of land weighs 5.2 pounds. About how many gallons of rain are there? Round your answer to the nearest tenth.

Copyright © Big Ideas Learning, LLC
All rights reserved.

Name____ Date__________

14.4 Enrichment and Extension

Sports Statistics and Proportions

In baseball, WHIP is a pitching statistic used to find the average number of walks and hits allowed per inning. It is calculated by adding the walks and hits that a pitcher allows and then dividing that by the number of innings pitched.

$$\text{WHIP} = \frac{\text{Walks} + \text{Hits}}{\text{Innings Pitched}}$$

In hockey, a goalie's save percentage is the number of saves divided by the number of shots on goal.

$$\text{Save percentage} = \frac{\text{Saves}}{\text{Shots on goal}}$$

1. A pitcher allowed 26 walks and 104 hits in 85 innings pitched. What is his WHIP? Round your answer to the nearest hundredth.

2. The pitcher in Exercise 1 wants to lower his WHIP to 1.3. How many more innings would he have to pitch without any walks or hits? Is this a reasonable goal? Explain your reasoning.

3. Another pitcher has a WHIP of 1.28 with 92 hits in 82 innings pitched. How many walks has he allowed? Explain how you used proportions to get your answer.

4. The pitcher from Exercise 3 wants to keep his WHIP under 1.3. How many more hits and walks can he allow in the next 9 innings? Is this a reasonable goal? Explain your reasoning.

5. A goalie has a save percentage of 0.914 after playing his first 6 games. He has made 191 saves. How many shots on goal has he faced?

6. Another goalie has a save percentage of 0.907 with 204 saves in his first 6 games. Your friend says that you could find their combined save percentage by adding 0.907 and 0.914 and then dividing by 2. Explain why this is incorrect. Then show how you would correctly calculate their combined save percentage. Round your answer to the nearest thousandth.

7. The goalie from Exercise 5 wants to finish the next game with an overall save percentage of 0.925. How many shots on goal would he have to face without allowing any goals? Based on the goalie's first 6 games, is it reasonable for him to achieve this goal in one game? Explain your reasoning.

Copyright © Big Ideas Learning, LLC
All rights reserved.

Name ______________________________ Date __________

14.4 Puzzle Time

Did You Hear About...

A	B	C	D	E	F
G	H	I	J	K	L
M	N	O	P	Q	R

Complete each exercise. Find the answer in the answer column. Write the word under the answer in the box containing the exercise letter.

44 THE
20 BECAUSE
18 IN
16 DAY
3.8 MINUTES
45 SECONDS
9 FOR
5.2 LUNCH
$12\frac{3}{4}$ TIME
$18\frac{3}{4}$ CAFETERIA

Use multiplication to solve the proportion.

A. $\frac{m}{8} = \frac{3}{4}$

B. $\frac{7}{9} = \frac{y}{18}$

C. $\frac{6}{13} = \frac{r}{39}$

D. $\frac{g}{48} = \frac{11}{12}$

E. $\frac{z}{24} = \frac{25}{32}$

F. $\frac{b}{21} = \frac{5}{7}$

G. $\frac{11}{27} = \frac{n}{18}$

H. $\frac{9}{28} = \frac{s}{42}$

Use the Cross Products Property to solve the proportion.

I. $\frac{c}{12} = \frac{5}{3}$

J. $\frac{9}{4} = \frac{x}{16}$

K. $\frac{7}{8} = \frac{14}{p}$

L. $\frac{12}{7} = \frac{36}{n}$

M. $\frac{k}{20} = \frac{13}{50}$

N. $\frac{15}{a} = \frac{25}{14}$

O. $\frac{6.6}{1.2} = \frac{w}{4.2}$

P. $\frac{1.6}{3.2} = \frac{2.8}{t}$

Q. $\frac{5}{\$13.75} = \frac{p}{\$24.75}$

R. $\frac{230 \text{ cal}}{30 \text{ min}} = \frac{345 \text{ cal}}{x \text{ min}}$

36 EVERY
14 CLOCK
5.6 BACK
15 THAT
6 THE
21 AT
23.1 WENT
8.4 IT
$13\frac{1}{2}$ SLOW
$7\frac{1}{3}$ WAS

Copyright © Big Ideas Learning, LLC
All rights reserved.

Activity 14.5 Start Thinking!
For use before Activity 14.5

Explain how judges use math to determine who receives the first place prize at a dance competition.

Activity 14.5 Warm Up
For use before Activity 14.5

Plot and label the points in a coordinate plane.

1. $A(2, -4)$
2. $B(-3, 1)$
3. $C(0, -3)$
4. $D(2, 0)$
5. $E(-1, -1)$
6. $F(-3, 0)$

Copyright © Big Ideas Learning, LLC
All rights reserved.

Lesson 14.5 Start Thinking!
For use before Lesson 14.5

What math concepts are related to auto racing?

Lesson 14.5 Warm Up
For use before Lesson 14.5

Find the missing distances. Use the table to complete a line graph for each distance. Which graph is steeper?

Time (seconds)	Distance 1 (feet)	Distance 2 (feet)
0	0	0
1	12	6
2		
3		
4		

Copyright © Big Ideas Learning, LLC
All rights reserved.

Name________________________________ Date__________

14.5 Practice A

Find the slope of the line.

1. 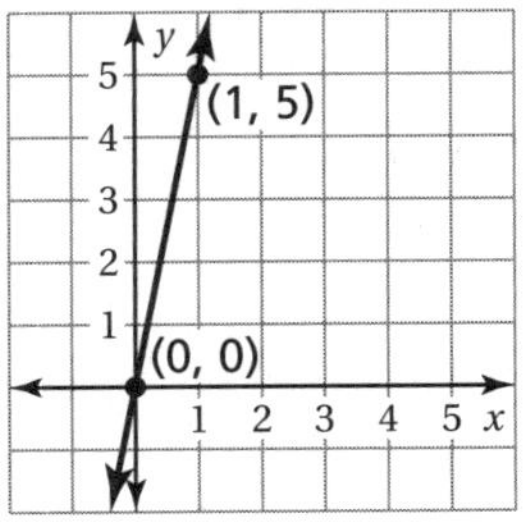

2. 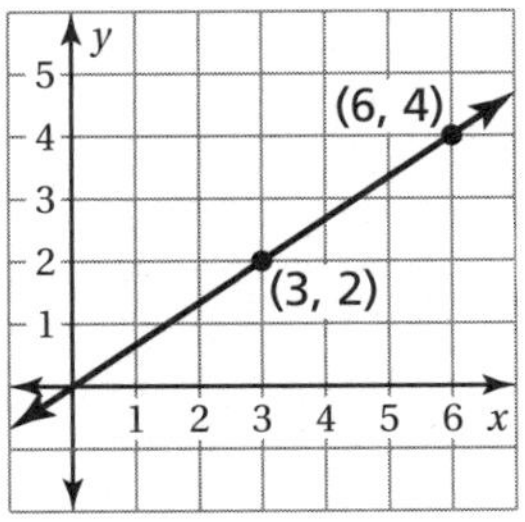

3. 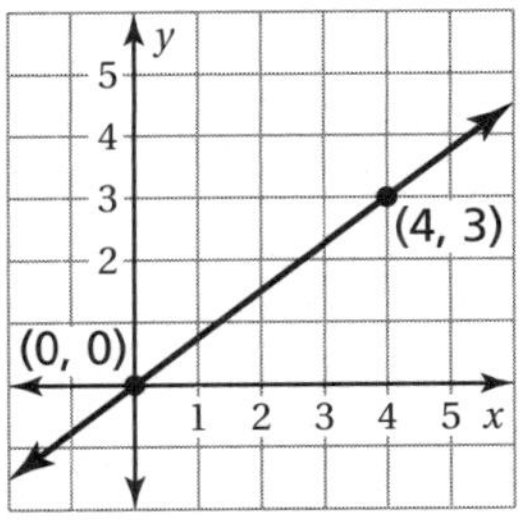

4. 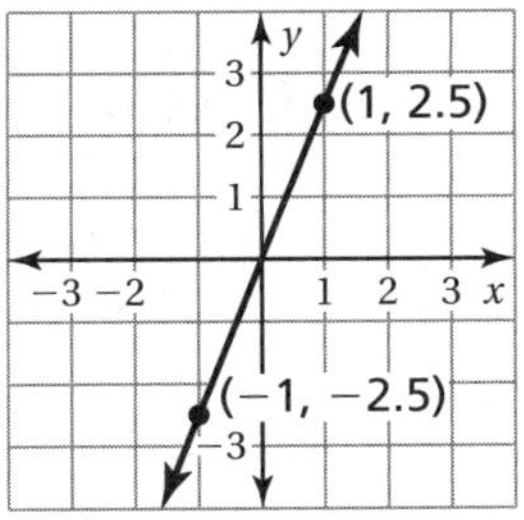

Graph the data. Then find and interpret the slope of the line through the points.

5.

Days, *x*	2	4	6	8
Pages, *y*	80	160	240	320

6.

Seconds, *x*	10	20	30	40
Feet, *y*	22	44	66	88

Graph the line that passes through the two points. Then find the slope of the line.

7. $(0, 0), (4, 3)$

8. $(-1, -2), (2, 4)$

9. $(-4, -1), (8, 2)$

10. The graph shows the amounts that you are collecting for selling calendars and boxes of greeting cards to raise money for the school band.

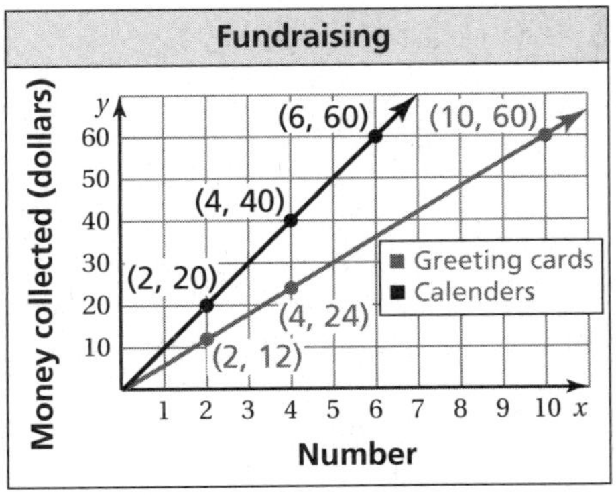

a. Compare the steepness of the lines. What does this mean in the context of the problem?

b. Find the slope of each line. What does each slope mean in the context of the problem?

c. How much more does it cost to buy 3 calendars than 4 boxes of greeting cards?

d. Find two different ways that you could collect exactly \$36.

Copyright © Big Ideas Learning, LLC
All rights reserved.

Name ________________________________ Date ________

14.5 Practice B

Find the slope of the line.

1.

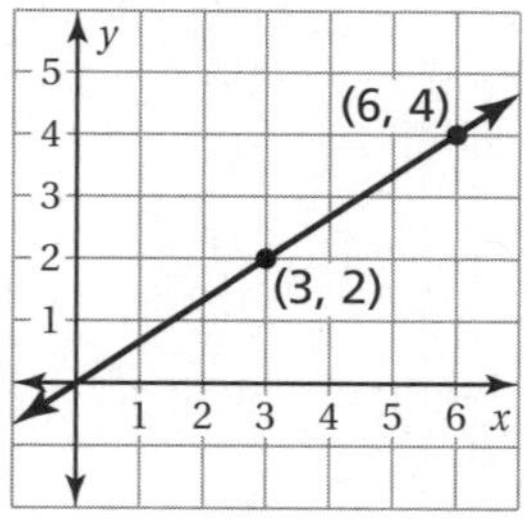

2.

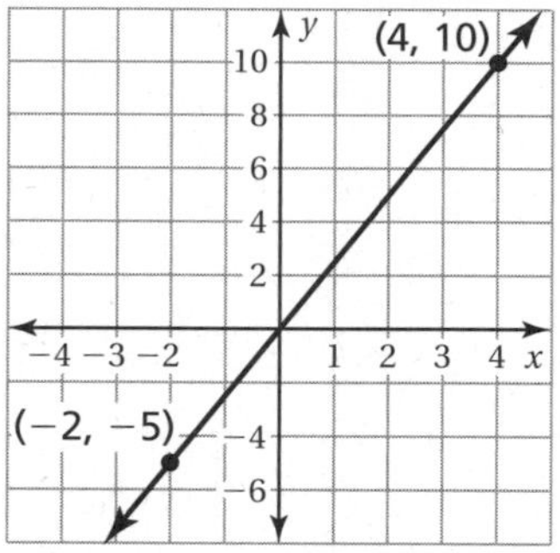

Which line has the greater slope? Explain your reasoning.

3.

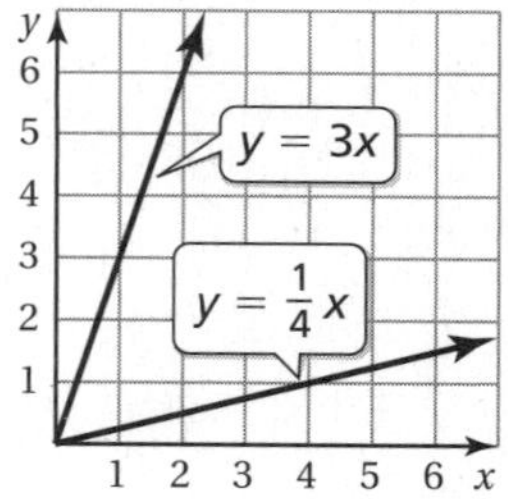

4.

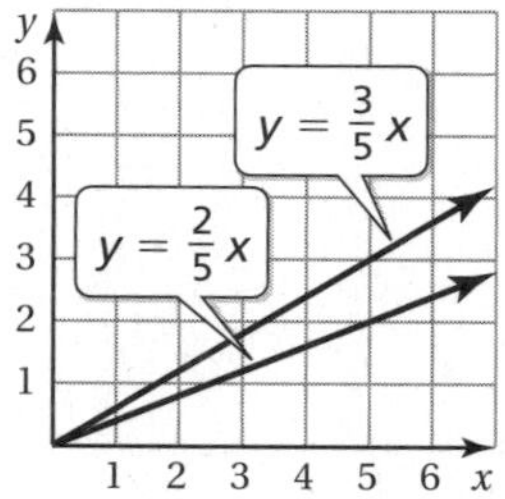

Graph the line that passes through the two points. Then find the slope of the line.

5. $(-2, -3), (3, 4.5)$

6. $(-8, -10), (-2, -2.5)$

7. Graph the line that passes through the points $(0, 1)$ and $(0, 4)$. Explain why the slope of this line is undefined.

8. You and a friend throw tennis balls up in the air at the same time. The table shows the height (in feet) of each ball.

Seconds, x	2	4	5
You, y	12	24	30
Friend, y	11	22	27.5

a. Graph the data on the same coordinate axes. Draw lines through the points. Label each graph.

b. Find the slope of the line for each tennis ball. What does each slope mean in the context of the problem?

c. Which ball is moving faster? How is this indicated in the slope?

d. Find the height of each ball 3.5 seconds after being thrown.

9. A line has a slope of $\frac{3}{5}$. It passes through the points $(5, 3)$ and $(x, 9)$. What is the value of x?

Copyright © Big Ideas Learning, LLC
All rights reserved.

Name________________________________ Date__________

14.5 Enrichment and Extension

Finding Slope Without Drawing a Graph

You can find the slope of a line through two points without drawing a graph.
To find the change in y without a graph, subtract the y-values of two points.
To find the change in x without a graph, subtract the x-values of two points.
Be sure to put the coordinates of the points in the same order when subtracting.

Example: Find the slope of the line that passes through the points $(2, 7)$ and $(3, 10)$.

Let (x_1, y_1) be $(2, 7)$ and (x_2, y_2) be $(3, 10)$.

$$\frac{\text{change in } y}{\text{change in } x} = \frac{y_2 - y_1}{x_2 - x_1}$$ Write coordinates in the same order.

$$= \frac{10 - 7}{3 - 2}$$ Substitute.

$$= \frac{3}{1} = 3$$ Simplify.

The slope of the line is 3.

Without graphing, find the slope of the line through the points.

1. $(2, 5), (5, 6)$

2. $(-1, -2), (-5, -7)$

3. $(-6, 0), (2, 8)$

4. $(4, 1), (-8, 10)$

5. In the example above, let (x_1, y_1) be $(3, 10)$ and (x_2, y_2) be $(2, 7)$. Would the slope still be the same? Explain your reasoning.

6. A line has a slope of $\frac{3}{2}$. It passes through the point $(4, 5)$. Give the coordinates of three other points on this line.

7. A line has a slope of $\frac{4}{7}$. It passes through the points $(2, -4)$ and $(-5, y)$. What is the value of y?

8. A line has a slope of 5. It passes through the points $(-5, -1)$ and $(x, 9)$. What is the value of x?

Copyright © Big Ideas Learning, LLC
All rights reserved.

Name ____ Date __________

14.5 Puzzle Time

What Is The Invisible Man's Favorite Drink?

Circle the letter of each correct answer in the boxes below. The circled letters will spell out the answer to the riddle.

Graph the line that passes through the two points. Then find the slope of the line.

1. $(0, 0), (6, 7)$ **2.** $(0, 0), (-3, -5)$ **3.** $(1, 2), (4, 8)$

4. $(2, 2), (5, 5)$ **5.** $(-4, -12), (2, 6)$ **6.** $(-9, -2), (18, 4)$

7. $(-6, -2), (6, 2)$ **8.** $(-2, -8), (5, 20)$ **9.** $(10, 12), (20, 24)$

10. $(-12, -8), (12, 8)$ **11.** $(-12, -2), (6, 1)$ **12.** $(8, 1), (24, 3)$

In Exercises 13 and 14, use the table below for the price of admission to a water park.

Water Park Admission				
Number of Persons	2	4	6	8
Child	\$46	\$92	\$138	\$184
Adult	\$62	\$124	\$186	\$248

13. Find the slope of the line for the price of a child's admission to the water park.

14. Find the slope of the line for the price of an adult's admission to the water park.

C	**A**	**E**	**R**	**V**	**L**	**A**	**T**	**D**	**P**	**G**	**I**	**O**	**N**	**E**	**R**
8	$\frac{1}{2}$	3	$\frac{2}{5}$	$\frac{1}{6}$	$\frac{3}{4}$	$\frac{1}{8}$	5	$\frac{1}{4}$	$\frac{2}{9}$	$\frac{6}{7}$	$\frac{5}{4}$	1	7	$\frac{1}{12}$	2
A	**M**	**K**	**T**	**L**	**E**	**D**	**I**	**M**	**A**	**R**	**I**	**T**	**L**	**K**	**S**
23	$\frac{7}{8}$	6	$\frac{7}{6}$	$\frac{3}{5}$	$\frac{1}{3}$	$\frac{2}{3}$	12	4	9	25	$\frac{5}{3}$	90	31	$\frac{6}{5}$	0

Copyright © Big Ideas Learning, LLC
All rights reserved.

Activity 14.6 Start Thinking!

For use before Activity 14.6

You made a sketch of the school mascot. Your sketch has a height of 2 inches and a width of 4 inches. Describe the proportions you would use to make a mural of your sketch on a wall in your school. What would be the dimensions of your mural?

Activity 14.6 Warm Up

For use before Activity 14.6

Plot the point in a coordinate plane. Describe the location of the point.

1. $A(-4, -2)$

2. $B(-2, -1)$

3. $C(0, 0)$

4. $D(2, 1)$

5. $E(4, 2)$

6. $F(6, 3)$

Copyright © Big Ideas Learning, LLC
All rights reserved.

Lesson 14.6 Start Thinking!

For use before Lesson 14.6

You receive \$20 every time you mow your neighbor's lawn. How many times do you need to mow the lawn so that you can buy a digital camera that costs \$189? Write an equation to represent the situation. Does the equation show direct variation? Why or why not?

Lesson 14.6 Warm Up

For use before Lesson 14.6

Graph the ordered pairs in a coordinate plane. Do you think that the graphs show that the quantities vary directly? Explain your reasoning.

1. $(0, 0), (1, 6), (2, 12), (3, 18)$

2. $(-3, 6), (4, 8), (-5, 10), (6, 12)$

3. $(4, 6), (8, 12), (12, 18), (18, 27)$

Copyright © Big Ideas Learning, LLC
All rights reserved.

Name____ Date__________

14.6 Practice A

Graph the ordered pairs in a coordinate plane. Do you think that graph shows that the quantities vary directly? Explain your reasoning.

1. $(-2, -2), (0, 0), (2, 2), (4, 4)$

2. $(-1, -4), (0, -1), (1, 2), (2, 5)$

Tell whether *x* and *y* show direct variation. Explain your reasoning. If so, find *k*.

3.

x	−1	0	1	2
y	2	0	2	4

4.

x	2	4	6	8
y	1	2	3	4

5. $y - 2 = 3x - 2$

6. $y + 3 = x$

7. $xy = 5$

8. The table shows the grams of fiber y for the grams of protein x. Graph the data. Tell whether x and y show direct variation. If so, write an equation that represents the line.

Grams of protein, x	3	6	9	12
Grams of fiber, y	2	4	6	8

The variables *x* and *y* vary directly. Use the values to find the constant of proportionality and write an equation that relates *x* and *y*.

9. $y = 6;\ x = 2$

10. $y = 15;\ x = 3$

11. $y = 40;\ x = 10$

12. To prepare an aquarium for use, you can clean it with a saltwater solution. The amount of salt varies directly with the volume of the water. The solution has 2 teaspoons of aquarium salt for every gallon of water.

    a. How many teaspoons of aquarium salt are needed for 5 gallons of water?

    b. Write an equation that relates x gallons of water to y teaspoons of salt.

    c. Use the equation to find the number of gallons of water to use for 12 teaspoons of salt.

13. The total cost of football tickets varies directly with the number of tickets purchased. Four tickets cost \$32. How many tickets can you buy for \$56?

14. One quart is equivalent to 0.95 liter.

    a. Write a direct variation equation that relates x quarts to y liters.

    b. Write a direct variation equation that relates x gallons to y liters.

    c. Write a direct variation equation that relates x liters to y quarts.

    d. What is the relationship between the values of k in the direct variation equations in parts (a) and (c)?

Copyright © Big Ideas Learning, LLC
All rights reserved.

Name ______________________ Date __________

14.6 Practice B

Graph the ordered pairs in a coordinate plane. Do you think that graph shows that the quantities vary directly? Explain your reasoning.

1. $(-3, -1), (3, 1), (6, 2), (9, 3)$

2. $\left(-2, -\frac{5}{2}\right), (4, 5), \left(6, \frac{15}{2}\right), (12, 15)$

Tell whether *x* and *y* show direct variation. Explain your reasoning. If so, find *k*.

3. $y - 2x = x$

4. $y - 5 = 5x + 2$

5.

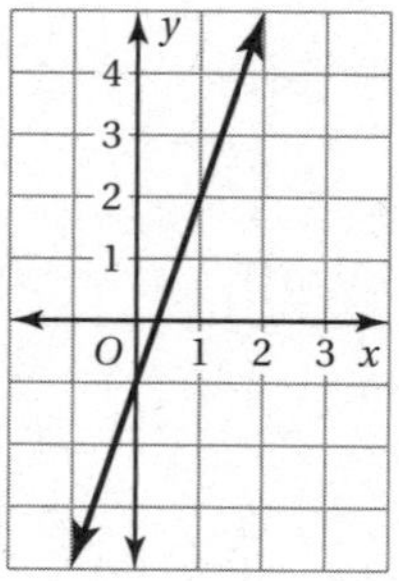

6.

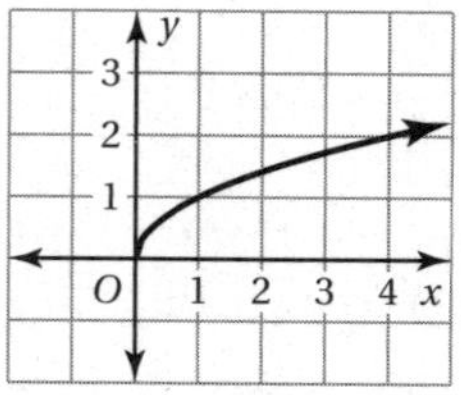

7. The percent y of correct answers on a test varies directly with the points x earned on the test.

Points earned, *x*	?	48	?	64	72	?
Percent, *y*	50	60	70	80	90	100

a. Copy and complete the table.

b. Write an equation that relates x and y.

c. Graph the points. Draw a line through the points. Does the graph confirm that x and y show direct variation? Explain your reasoning.

The variables *x* and *y* vary directly. Use the values to find the constant of proportionality and write an equation that relates *x* and *y*.

8. $y = 36;\ x = 18$

9. $y = 51;\ x = 34$

10. $y = 55;\ x = 10$

11. The table shows the heights of the vertical supports for two skateboard ramps. Which ramp(s) show direct variation between distance and height?

Distance from base (feet)	1	2	3	4
Height for your ramp (feet)	1	1.5	2	2.5
Height for friend's ramp (feet)	0.5	1	1.5	2

12. Does the graph of every direct variation equation pass through the origin? Is every relationship whose graph passes through the origin direct variation? Explain your reasoning.

Copyright © Big Ideas Learning, LLC
All rights reserved.

Name____ Date__________

14.6 Enrichment and Extension

More on Direct Variation

1. A store's profit P varies directly with the number of items n that they purchase from a warehouse. Greg's Gadgets buys 75 Wacky Widgets from the Gadget Supply Warehouse. The store's profit for this purchase is −\$190.50.

   a. Write a direct variation equation. Then, explain what the equation means.

   b. Explain why the profit is negative. Predict what a graph of the equation will look like.

   c. Fill in the table and make a graph of the data. Then compare the graph with your prediction.

Items purchased, *n*	0	5	10	15	20
Profit, *P*					

2. The volume V of blood that your heart pumps varies directly with your pulse rate p. Tomiqua's heart pumps 2.6 liters of blood in one minute when she is at her resting heart rate of 52 beats per minute.

   a. Write a direct variation equation. Then, explain what the equation means.

   b. After Tomiqua has been running on a treadmill, she measures her pulse for 10 seconds and gets 15 beats. How much blood is her heart pumping per minute?

3. The weight of an object on the moon varies directly with its weight on Earth. For example, the Apollo 13 Command Module weighed approximately 6 tons on Earth and about 1 ton on the moon.

   a. Write a direct variation equation. Then, explain what the equation means.

   b. Find your approximate weight on the moon.

Copyright © Big Ideas Learning, LLC
All rights reserved.

Name ______________________________ Date __________

14.6 Puzzle Time

How Do Bees Get To School?

For each exercise, circle the letter in the columns under Yes or No to indicate the correct answer. The circled letters will spell the answer to the riddle.

	Yes	No
1.	T	S
2.	H	T
3.	A	E
4.	Y	R
5.	M	T
6.	E	A
7.	L	K
8.	E	R
9.	T	S
10.	O	H
11.	E	K
12.	A	B
13.	U	N
14.	Z	T
15.	Y	Z

Tell whether *x* and *y* show direct variation.

1. $(1, 1), (2, 2), (3, 3), (4, 4)$
2. $(1, 3), (2, 6), (3, 9), (4, 12)$
3. $(-2, 3), (0, 0), (2, 3), (4, 12)$
4. $(-4, -1), (4, 1), (8, 2), (12, 3)$
5. $(-1, -1), (0, 1), (1, 3), (2, 5)$
6. $(1, 1), (2, 4), (3, 9), (4, 16)$
7. $(1, 3), (2, 3), (3, 3), (4, 3)$
8. $y = 12x$
9. $y = \frac{1}{7}x$
10. $y = 4x^2$
11. $y = -2x$
12. $y = 7x - 3$
13. $x = 5y$
14. $6 = \frac{x}{y}$
15. $x^3 = 12y$

Copyright © Big Ideas Learning, LLC
All rights reserved.

Name________________________________ Date __________

Chapter 14

Technology Connection

For use after Section 14.5

Exploring Slope

Dynamic geometry software allows you to explore the slope of a line or line segment. The directions below might not correspond exactly to your software, but the same actions can be performed using similar commands.

Open a **New Sketch**. Use the **Segment** tool on the left side of the screen to draw a short segment that slants upward.

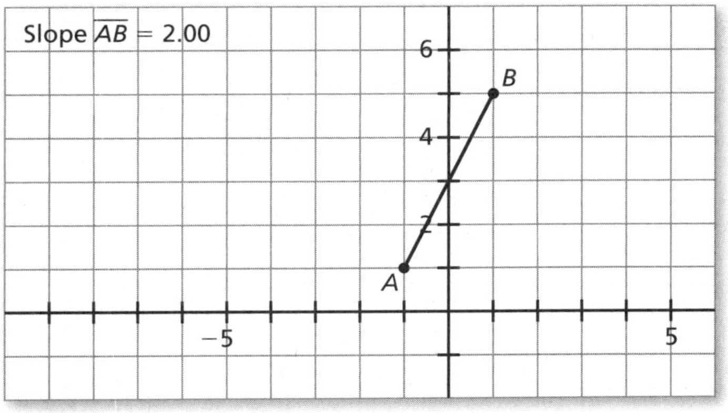

Click on the middle of the segment you made. In the **Measure** menu, select **Slope**. A grid and the slope of the segment will appear on the screen. In the **Graph** menu, select **Snap Points**. This will make it so that the segment will always have integer coefficients.

Move the endpoints of the segment you made so that it has the given slope. Sketch the line segment on a piece of paper and label the slope.

1. 1 **2.** $\frac{1}{2}$ **3.** $\frac{2}{3}$ **4.** $-\frac{4}{5}$

5. 0 **6.** −0.75 **7.** $1.\overline{3}$ **8.** −2.5

9. Which line segments are the steepest? Which are the least steep? Describe their slopes.

10. What is the slope of a horizontal line segment?

11. What happens to the slope when you make the line segment vertical? Why?

12. Can you move the line segment without changing the slope? How?

Copyright © Big Ideas Learning, LLC
All rights reserved.

Chapter 15

Copyright © Big Ideas Learning, LLC
All rights reserved.

Name________________________________ Date__________

Chapter 15 Percents

Dear Family,

Family vacations provide a relaxing break from the usual routine. Whether you visit family, go camping, or plan a "stay-cation" at home, you usually need to make a budget for your expenses.

Budgeting for a family vacation may take some research. The rates advertised for hotel rooms usually do not include taxes or resort fees. When dining out, the tax and tip can add quite a bit to the total bill. Vehicle rentals often have extra charges for taxes, insurance, fuel, and mileage that are not usually included in advertised rates. All of these "hidden" costs can increase the cost of your vacation significantly.

Ask your student to help create a budget for your next family vacation. Have him or her research common fees, taxes, and tips for each part of your plan. For example:

- Hotels sometimes have multiple tax rates—for state, county, and local sales taxes, and sometimes even an extra "room tax." Find out what the tax rates are and determine how much this will add to the room charges. Are there other fees, such as resort fees or parking charges?
- Employees in restaurants and hotels usually earn most of their money from tips, so find out who usually receives a tip and what amount is appropriate.
- Rental cars can have many taxes and fees—sometimes these cost as much as the rental fee itself. Try to find out each rate or fee so you can plan accordingly.

Work with your student to make a budget that includes the extra charges for each item. You might want to include other categories in your budget—such as categories for shopping, entertainment, fuel, and parking. Having your student help with the budgeting will involve him or her in planning the vacation—and help your student realize you can't always afford everything you want to do.

Don't forget to write!

Copyright © Big Ideas Learning, LLC
All rights reserved.

Nombre ____ Fecha ________

Capítulo 15 Porcentajes

Estimada Familia:

Las vacaciones familiares ofrecen un descanso relajante de la rutina cotidiana. Ya sea que se visite a la familia, se vaya de campamento o se quede en casa descansando, generalmente se necesita hacer un presupuesto para los gastos.

Hacer el presupuesto para una vacación familiar puede demandar algo de investigación. Por lo general, las tarifas publicadas de habitaciones de hotel no incluyen impuestos ni costos del complejo.

Al salir a cenar, el impuesto y la propina aumentan de manera regular la cuenta total. Los alquileres de vehículos a menuda cobran extra por impuestos, seguro, gasolina y millaje, lo que generalmente no está incluido en las tarifas publicadas. Todos estos costos "ocultos" pueden aumentar el costo de las vacaciones de manera significativa.

Pida a su estudiante que lo ayude a crear un presupuesto para sus próximas vacaciones familiares. Haga que investigue acerca de precios, impuestos y propinas comunes para cada parte de su plan. Por ejemplo:

- Los hoteles a veces cobran impuestos múltiples—impuestos de ventas estatales, nacionales y locales, y a veces incluso cobran un "impuesto por habitación" adicional. Averigüen las tarifas de impuestos y determinen en cuánto aumentará la cuenta de la habitación. ¿Hay otros gastos, como los costos del complejo o del estacionamiento?
- Los empleados de restaurantes y hoteles generalmente ganan la mayoría de su dinero en propinas, así que averigüen quién recibe generalmente una propina y qué cantidad es la apropiada.
- Los autos de alquiler pueden incluir muchos impuestos y gastos—a veces estos cuestan tanto como el alquiler del coche mismo. Intenten encontrar cada tarifa o gasto para que puedan planear como corresponde.

Trabaje con su estudiante para hacer un presupuesto que incluya los gastos extras de cada artículo. Querrán incluir otras categorías en su presupuesto—como por ejemplo: compras, entretenimiento, gasolina y estacionamiento. El permitir que su estudiante ayude con el presupuesto, lo animará a que planee las vacaciones—y ayude a su estudiante a darse cuenta que no siempre se puede conseguir todo lo que se desea.

¡No se olviden de escribir!

Copyright © Big Ideas Learning, LLC
All rights reserved.

Activity 15.1 Start Thinking!

For use before Activity 15.1

Study the decimal and percent equivalents below. What do you notice?

0.35 = 35%	0.27 = 27%	0.03 = 3%
0.2 = 20%	0.5 = 50%	1.38 = 138%
2.1 = 210%	6 = 600%	0.886 = 88.6%

Andrew says that 0.98 = 9.8%. Is he correct? Why or why not?

Activity 15.1 Warm Up

For use before Activity 15.1

Write the fraction represented by the model.

1.

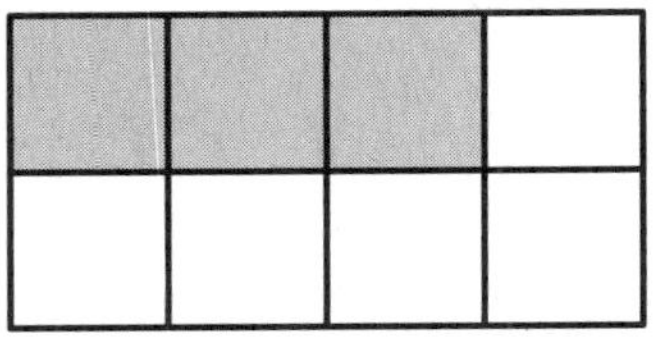

2.

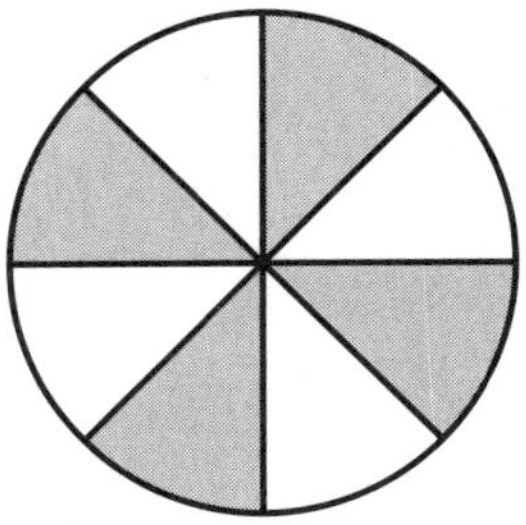

3.

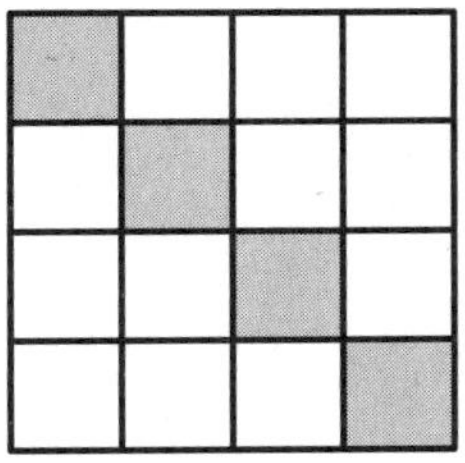

4.

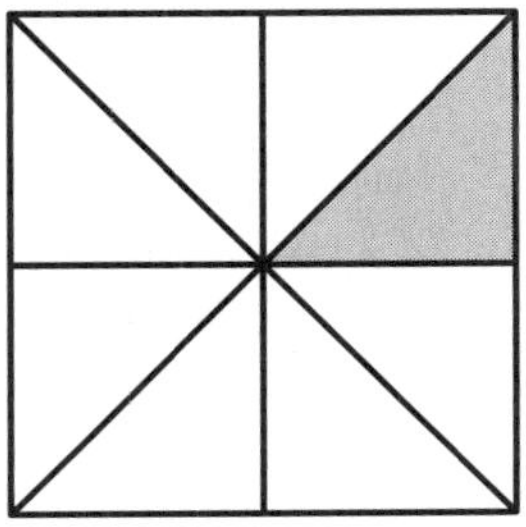

5.

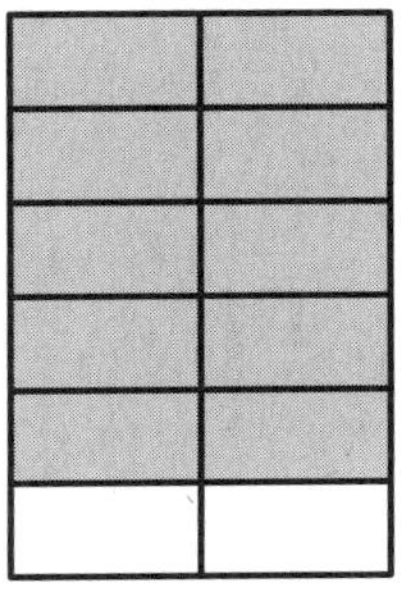

6.

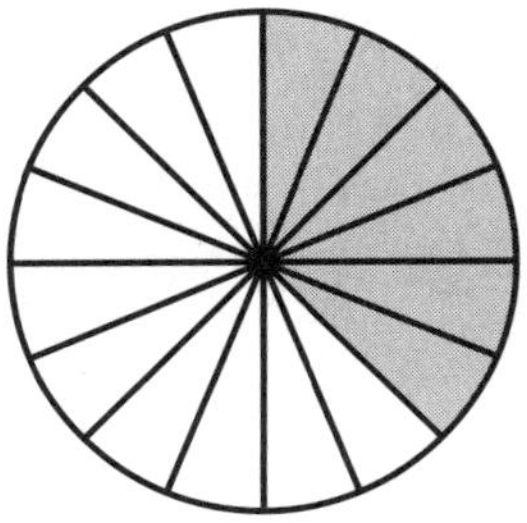

Copyright © Big Ideas Learning, LLC
All rights reserved.

Lesson 15.1 Start Thinking!
For use before Lesson 15.1

Describe a situation in sports when decimals and percents are used interchangeably.

Lesson 15.1 Warm Up
For use before Lesson 15.1

Write the percent as a decimal.

1. 19% **2.** 2% **3.** 7.5%

Write the decimal as a percent.

4. 0.89 **5.** 0.54 **6.** 0.1

Copyright © Big Ideas Learning, LLC
All rights reserved.

Name______________________________ Date__________

15.1 Practice A

Write the percent as a decimal.

1. 81% **2.** 78% **3.** 5% **4.** 8%

5. 40% **6.** 60% **7.** 23.7% **8.** 16.75%

9. 150% **10.** 210% **11.** 186% **12.** 416%

13. 100.8% **14.** 5.17% **15.** 0.4% **16.** 0.04%

17. Describe and correct the error in writing 1.475% as a decimal.

✗ 1.475% = 1.475% = 147.5

Write the decimal as a percent.

18. 0.66 **19.** 0.32 **20.** 0.51 **21.** 0.97

22. 0.01 **23.** 0.04 **24.** 0.312 **25.** 0.468

26. 0.5 **27.** 1.2 **28.** 1.08 **29.** 1.16

30. 0.003 **31.** 0.025 **32.** 0.0245 **33.** 2.025

34. Describe and correct the error in writing 1.8 as a percent.

✗ 1.8 = 1.8 = 18%

35. Fifty-four percent of the students in your class have moved at least one time. Write this percent as a decimal.

36. Only 0.15 of the total number of vehicles in your school parking lot are buses. What percent of the vehicles are buses?

37. You spent 0.88 of your allowance this week. What percent of your allowance did you spend?

38. On a history test, you get 86 out of a possible 100 points. Write a decimal and a percent that represent a score of 86 out of 100.

39. Of the fluids that you drink on a typical day, $\frac{1}{10}$ is milk and 50% is water. How many times more water do you drink than milk?

Write the percent as a fraction in simplest form and as a decimal.

40. 21% **41.** 75% **42.** 64% **43.** 85%

Copyright © Big Ideas Learning, LLC
All rights reserved.

Name ______________________________ Date __________

15.1 Practice B

Write the decimal as a percent.

1. 0.54 **2.** 0.37 **3.** 0.222 **4.** 0.929

5. 1.4 **6.** 2.5 **7.** 20 **8.** 0.005

Write the percent as a fraction in simplest form and as a decimal.

9. 68% **10.** 9% **11.** 55% **12.** 26%

13. 42.4% **14.** 73.6% **15.** 31.25% **16.** 44.65%

17. About 36% of the students at a middle school are seventh graders. What percent are *not* in seventh grade?

18. The percents of three types of tickets collected at the gate for a high school football game are shown.

Ticket type	Student	Adult	Senior (65 and older)
Percent	48%	28%	14%

a. Write the percents as decimals and as fractions.

b. There is one other type of ticket that is not shown. It is a ticket for a child under 5. What percent of the tickets were of this type?

c. Make a bar graph to represent the percents for all four ticket types.

19. Students in an after-school enrichment program chose one of five subject areas.

a. What percent chose English or reading?

b. What percent chose English or history?

c. How many times more students chose English than reading?

d. What percent chose math? Write the percent as a decimal.

Subject Chosen

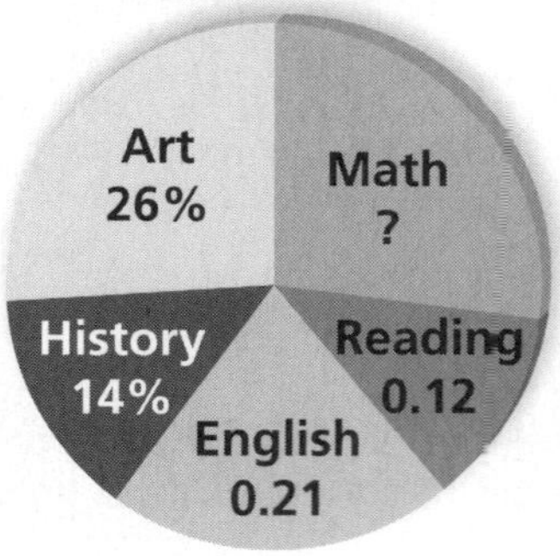

20. At one school, half of the students live within 1 mile, 78% live within 2 miles, and 0.1 of the students live between 2 and 3 miles from the school.

Make a table to show the percent of students who live at each distance from the school.

a. within 1 mile **b.** between 1 and 2 miles

c. between 2 and 3 miles **d.** more than 3 miles

Copyright © Big Ideas Learning, LLC
All rights reserved.

Name____ Date__________

15.1 Enrichment and Extension

Playing Checkers

You set up a game of checkers as shown. Write a decimal for each percent you find.

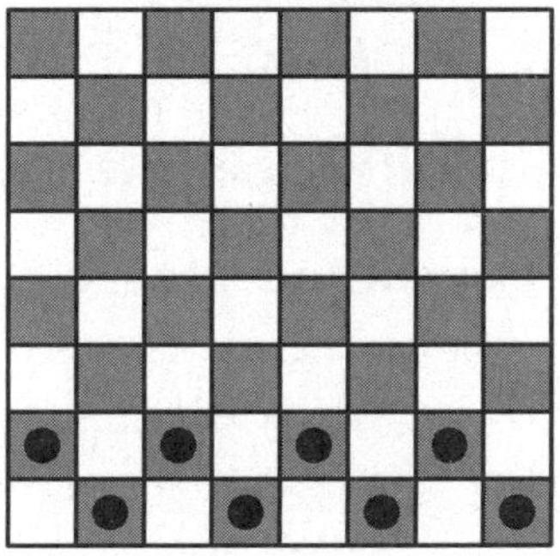

1. What percent of the checker board is shaded?

2. What percent of the checker board is *not* shaded?

3. Your friend arrives and places pieces appropriately on the board.

   a. What percent of the checker board squares have pieces?

   b. What percent of the checker board squares do *not* have pieces?

   c. What percent of the shaded checker board squares have pieces?

   d. What percent of the shaded checker board squares do *not* have pieces?

4. At one point during the game, 4 of your pieces have been removed and 5 of your friend's pieces have been removed.

   a. What percent of the original number of pieces have been removed?

   b. What percent of the original number of pieces are remaining?

5. The checker piece near the middle of the board can move diagonally in any direction between shaded squares.

   a. On how many different squares can the piece be located at the end of on turn?

   b. On how many different squares can the piece be located at the end of two turns? What percent of the shaded checker board is this? Write this percent as a decimal.

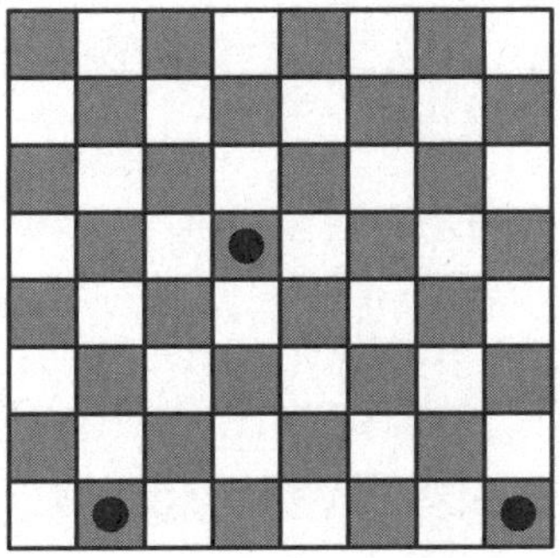

Copyright © Big Ideas Learning, LLC
All rights reserved.

Name ______________________________ Date __________

15.1 Puzzle Time

Why Are Math Assignments Like The Water That Is Found On The Ground In The Early Morning?

Write the letter of each answer in the box containing the exercise number.

Write the percent as a decimal.

1. 67% **2.** 44%

3. 29.6% **4.** 46.3%

5. 22% **6.** 8%

7. 58.74% **8.** 80.14%

9. 277% **10.** 106%

11. 0.05% **12.** 0.045%

Write the decimal as a percent.

13. 0.85 **14.** 0.41

15. 0.98 **16.** 0.657

17. 0.77 **18.** 0.51

19. 0.376 **20.** 0.239

21. 2.57 **22.** 0.0482

Answers for Exercises 1–12

T. 2.77 **E.** 0.00045

S. 0.296 **O.** 0.22

B. 0.67 **H.** 1.06

R. 0.8014 **P.** 0.5874

O. 0.463 **D.** 0.44

E. 0.0005 **U.** 0.08

Answers for Exercises 13–22

B. 37.6% **P.** 65.7%

S. 4.82% **A.** 23.9%

T. 85% **E.** 98%

U. 257% **D.** 51%

O. 77% **E.** 41%

19	17	9	10		20	8	12		3	6	7	16	5	22	15	2		13	4		1	11		18	21	14

Copyright © Big Ideas Learning, LLC
All rights reserved.

Activity 15.2 Start Thinking!

For use before Activity 15.2

What are two methods that you can use to compare two fractions?

Use one of the methods you described to tell whether $\frac{21}{40}$ or $\frac{7}{12}$ is greater. Which method did you use? Why?

Activity 15.2 Warm Up

For use before Activity 15.2

Order the decimals from least to greatest.

1. 1.4, 1.44, 4.1, 4.3, 3.3

2. 0.5, 1.75, 0.7, 1.57, 1.5

3. 0.44, 0.02, 0.2, 0.04, 0.24

4. 6.6, 0.6, 6.63, 6.06, 0.06

5. 0.2, 0.17, 0.02, 0.72, 2.27

6. 8.1, 0.81, 0.18, 1.8, 1.88

Copyright © Big Ideas Learning, LLC
All rights reserved.

Lesson 15.2 Start Thinking!
For use before Lesson 15.2

Design a survey with three yes or no questions to ask the students in your class. Take turns in class asking the survey questions. Record your results in a table like the one below. You can use fractions, percents, or decimals. Which of your questions had the greatest percent of students answering yes?

Question	Yes	No
Do you play a sport?	$\frac{19}{28}$	$\frac{9}{28}$
Do you play a musical instrument?	20%	80%
Do you have any siblings?	0.85	0.15

Lesson 15.2 Warm Up
For use before Lesson 15.2

Tell which number is greater.

1. $\frac{5}{13}$, 45%

2. 30%, 0.4

3. $\frac{9}{17}$, 50%

4. 0.6, $\frac{4}{7}$

5. $\frac{2}{5}$, 0.225

6. 240%, 0.24

Copyright © Big Ideas Learning, LLC
All rights reserved.

Name____ Date __________

15.2 Practice A

Tell which number is greater.

1. $\frac{3}{4}$, 70% **2.** $\frac{1}{2}$, 0.54 **3.** 0.21, 21% **4.** $\frac{2}{3}$, 66%

5. 0.482, 49% **6.** 16%, 0.108 **7.** $\frac{12}{25}$, 48% **8.** $\frac{1}{10}$, 12%

9. 1.2, 11% **10.** 58%, $\frac{31}{50}$ **11.** 5020%, $50\frac{1}{4}$ **12.** 12.25%, $\frac{1}{8}$

13. Describe and correct the error in comparing 0.7% and $\frac{17}{25}$.

> X $\frac{17}{25} \overset{\times 4}{=} \frac{68}{100} = 0.68\%$ (×4)
>
> 0.7% is greater than 0.68%, so 0.7% is the greater number.

Use a number line to order the numbers from least to greatest.

14. 0.64, $\frac{13}{20}$, 63% **15.** 45%, 0.46, $\frac{11}{25}$

16. 0.12, $\frac{1}{8}$, 0.135, 13% **17.** $\frac{15}{16}$, 90%, 0.925, $\frac{7}{8}$, 0.93

18. $3\frac{2}{3}$, 362%, 3.66, $3\frac{3}{5}$, 36 **19.** 0.3, 27.3%, $\frac{11}{40}$, 28%, 0.27

20. You use 8 fluid ounces of fruit juice in a recipe to make 64 fluid ounces of fruit punch. A fruit punch you can buy at the store has 10% real fruit juice. Which has a higher percent of fruit juice?

21. While shooting baskets at a basketball hoop, you make 36 out of 80 shots. Your friend makes 43.75% of the shots. Who made a higher percent?

22. To earn a bonus in a video game, you must find at least 60% of the hidden gems. You find 25 out of 40 gems. Do you get the bonus? Explain.

23. The table shows the portion of students at a middle school that are in each grade. Order the grades from the least to the greatest number of students.

Grade	6	7	8
Portion of students	$33\frac{1}{3}\%$	0.3125	$\frac{17}{48}$

Copyright © Big Ideas Learning, LLC
All rights reserved.

Name ____ Date __________

15.2 Practice B

Tell which number is greater.

1. $\frac{1}{4}$, 22% **2.** $\frac{5}{9}$, 55% **3.** 3.2, 32% **4.** 99.9%, 1

Use a number line to order the numbers from least to greatest.

5. $\frac{1}{3}$, 0.3, 33%, $\frac{8}{25}$, 33.6% **6.** 210%, 2.2, $2.\overline{2}$, $\frac{43}{20}$

Tell which letter shows the graph of the number.

7. 0.884 **8.** $\frac{8}{9}$ **9.** $\frac{22}{25}$ **10.** 0.89

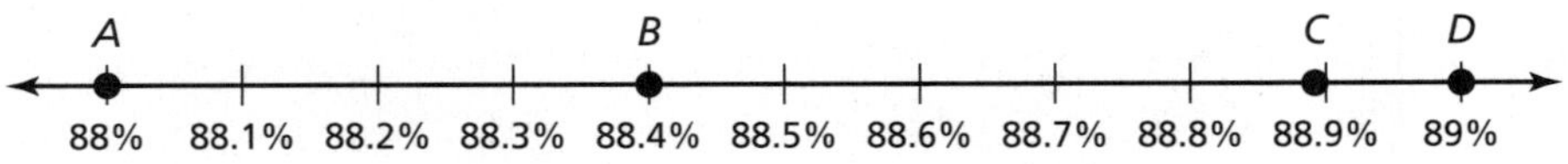

11. Describe a process that you can use to find a decimal whose value is between 31% and 32%.

12. Is 6 centimeters greater than 5% of a meter? Explain.

13. Does 6% of a pound weigh more than an ounce? Explain.

14. Order the periods of time from least to greatest.

1% of an hour $\frac{2}{3}$ of a minute 0.0004 of a day

15. The table shows the portions of the U.S. population that lived in Florida in certain years.

Year	1860	1910	1960	2010
Portion of U.S. Population in Florida	0.45%	0.0082	$\frac{1}{36}$	$\frac{1}{16}$

a. Order the portions from least to greatest.

b. Since 1860, how has the population of Florida increased compared to the population of the United States? Why do you think this happened?

c. Do you think this will always happen? Explain your reasoning.

16. Arsenic is toxic to humans. The greatest amount of arsenic that is allowed in drinking water is 10 parts per billion. A test shows that a source of drinking water contains 0.000002% arsenic. Is this an allowable amount? Explain.

Copyright © Big Ideas Learning, LLC
All rights reserved.

Name_______________________________ Date__________

15.2 Enrichment and Extension

Treasure Map

You want to find the treasure shown on the map. To get to the treasure, start at 1% and move to 100%. The only way you can move on the map is if a number is greater than the number you are currently on. List the numbers in order that make up your path to the treasure.

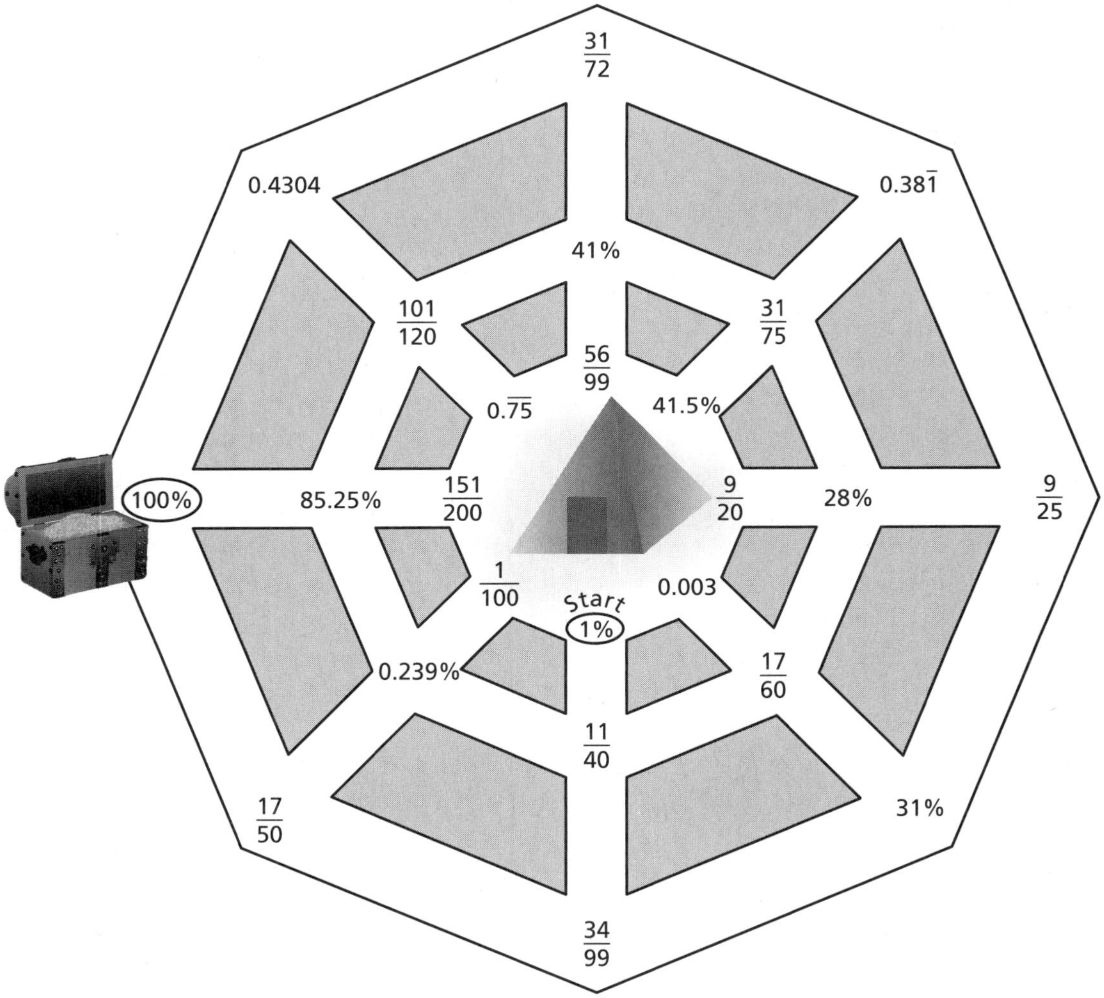

Copyright © Big Ideas Learning, LLC
All rights reserved.

Name ______________________________ Date __________

15.2 Puzzle Time

Why Did The Peach Need To Hire An Attorney?

Write the letter of each answer in the box containing the exercise number.

Tell which number is greater.

1. $\frac{17}{20}$, 95%

2. 30%, 0.03

3. $\frac{27}{50}$, 27%

4. $\frac{14}{25}$, 60%

5. 0.097, 97%

6. 65%, 0.56

7. 80%, $\frac{7}{8}$

8. 0.13, 13.5%

9. 11%, $\frac{3}{25}$

10. 150%, 0.15

11. $\frac{1}{4}$, 20%

12. 66%, $\frac{2}{3}$

Tell which number is the greatest.

13. 28%, $\frac{3}{5}$, 0.31

14. 79%, 0.52, $\frac{17}{20}$

15. $\frac{17}{25}$, 0.81, $\frac{5}{8}$, 64%

16. 0.14%, $\frac{7}{20}$, 0.014, 1.4

Answers for Exercises 1–12

W. $\frac{3}{25}$	**D.** $\frac{14}{25}$	**G.** 80%
E. 66%	**R.** $\frac{1}{4}$	**L.** $\frac{27}{50}$
Y. 97%	**I.** 20%	**E.** 11%
A. 95%	**T.** 150%	**C.** 0.15
J. $\frac{2}{3}$	**N.** $\frac{7}{8}$	**O.** 27%
I. 60%	**S.** 30%	**K.** 0.56
U. 13.5%	**M.** 0.03	**A.** 65%
P. 0.097	**B.** $\frac{17}{20}$	**F.** 0.13

Answers for Exercises 13–16

E. $\frac{5}{8}$	**D** 0.014	**T.** 0.81
W. 0.52	**A.** $\frac{3}{5}$	**S.** 0.31
M. $\frac{17}{20}$	**Y.** 79%	**I.** 1.4
M. 28%	**R.** 0.14%	**B.** 64%

4	15		9	13	2		10	11	8	3	5		16	7		6		12	1	14

Copyright © Big Ideas Learning, LLC
All rights reserved.

Activity 15.3 Start Thinking!

For use before Activity 15.3

Discuss with a partner how percents are used in a real-life situation.

Activity 15.3 Warm Up

For use before Activity 15.3

Estimate the sum or difference.

1. $162 + 98$ **2.** $148 - (-69)$ **3.** $-239 + 102$

Estimate the product or quotient.

4. $32(-43)$ **5.** $\frac{-187}{12}$ **6.** $(-49)(-12)$

Copyright © Big Ideas Learning, LLC
All rights reserved.

Lesson 15.3 Start Thinking!

For use before Lesson 15.3

At a soccer game, your team scored 6 goals during regular play and 4 goals on penalty kicks. Use a model to show the percent of goals that were not scored by penalty kicks.

Lesson 15.3 Warm Up

For use before Lesson 15.3

Use a model to estimate the answer to the question. Use a ratio table to check your answer.

1. What number is 20% of 60?

2. 35 is what percent of 70?

3. 51 is 102% of what number?

4. What number is 80% of 130?

5. 72 is what percent of 100?

6. 44 is 55% of what number?

Copyright © Big Ideas Learning, LLC
All rights reserved.

Name_______________________________________ Date__________

15.3 Practice A

Use a model to estimate the answer to the question. Use a ratio table to check your answer.

1. What number is 20% of 40?
2. 12 is what percent of 50?
3. 42 is 60% of what number?
4. What number is 150% of 92?

Write and solve a proportion to answer the question.

5. 40% of what number is 15?
6. 24 is 0.6% of what number?
7. What percent of 75 is 27?
8. 17 is what percent of 68?
9. Of the 60 seeds that you plant, 80% germinate. How many seeds germinate?
10. You are charged 6% sales tax. You purchase a new bicycle and pay $27 in sales tax. What is the purchase price of the bicycle?

Write and solve a proportion to answer the question.

11. 0.2 is what percent of 16?
12. 19.6 is 24.5% of what number?
13. $\frac{3}{5}$ is 30% of what number?
14. What number is 45% of $\frac{5}{9}$?
15. You are making 28 name badges for a committee. You complete 75% of these on Monday. How many do you have left to complete on Tuesday?
16. You and your friend are selling tickets for the orchestra concert. On Thursday, you sold 15 tickets and your friend sold 10 tickets.
    a. What percent of the tickets sold on Thursday did you sell?
    b. On Friday, you sold 9 tickets and your friend sold 16 tickets. What percent of the tickets sold on Friday did you sell?
    c. What percent of the total tickets sold on Thursday and Friday did you sell?

Copyright © Big Ideas Learning, LLC
All rights reserved.

Name ______________________________ Date __________

15.3 Practice B

Write and solve a proportion to answer the question.

1. 55% of what number is 33?

2. What percent of 120 is 42?

3. 36 is 0.8% of what number?

4. 48 is what percent of 64?

5. Of the 360 runners at a 5-kilometer race, 20% are in the 35–39 age bracket. How many runners at the 5-kilometer race are in the 35–39 age bracket?

6. You pay $3.69 for a gallon of gasoline. This is 90% of the price of a gallon of gasoline one year ago. What was the price of a gallon of gasoline one year ago?

7. Describe and correct the error in using the percent proportion to answer the question below.

   "6 is 6.25% of what number?"

   ✗ $\frac{a}{w} = \frac{p}{100}$

   $\frac{6}{w} = \frac{0.0625}{100}$

   $w = 9600$

Write and solve a proportion to answer the question.

8. $\frac{7}{8}$ is 70% of what number?

9. 7.2 is 250% of what number?

10. What number is 72% of $\frac{3}{8}$?

11. 1.4 is what percent of 1.12?

12. You earn a score of 86.8 on a standardized exam. Your score is 140% higher than your friend's score on the standardized exam. What is your friend's score?

13. 80% of a number is x. What is 40% of the number?

14. Answer each question.

    a. What is 35% of $90x$?

    b. What percent of $16x$ is $9x$?

Copyright © Big Ideas Learning, LLC
All rights reserved.

Name______________________________ Date__________

15.3 Enrichment and Extension

Proportions and Similar Triangles

Triangles that have the same shape but not necessarily the same size are called **similar triangles**. Two triangles are similar when corresponding side lengths are proportional and corresponding angles are congruent.

Example: The triangles are similar. Find *x*.

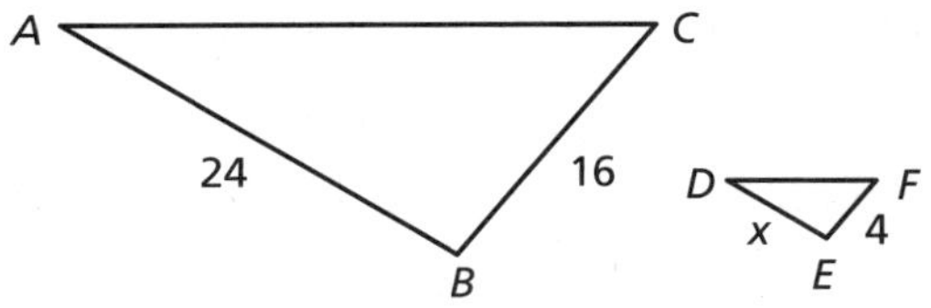

Triangle *ABC* Triangle *DEF*

$\frac{AB}{BC} = \frac{DE}{EF}$ Set up a proportion.

$\frac{24}{16} = \frac{x}{4}$ Substitute.

$24 \bullet 4 = 16 \bullet x$ Cross multiply.

$96 = 16x$ Simplify.

$6 = x$ Divide each side by 16.

So, $x = 6$.

The triangles are similar. Find *x*.

1.

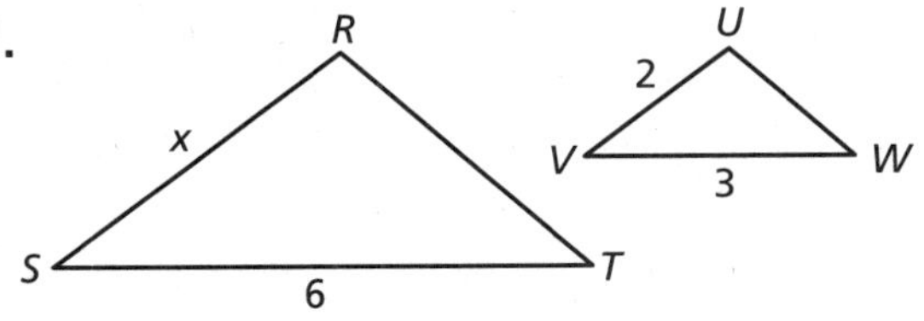

2.

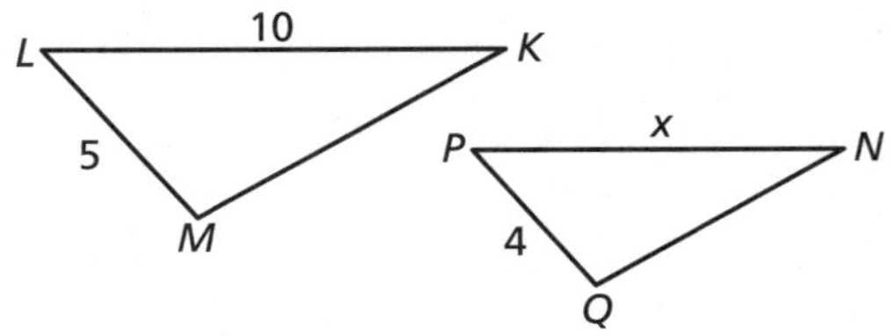

3.

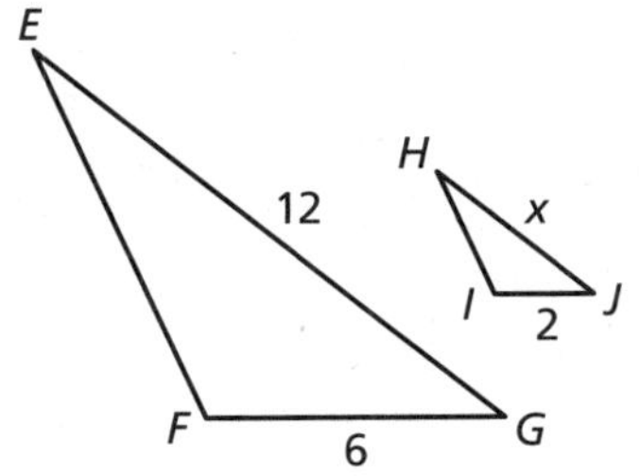

4.

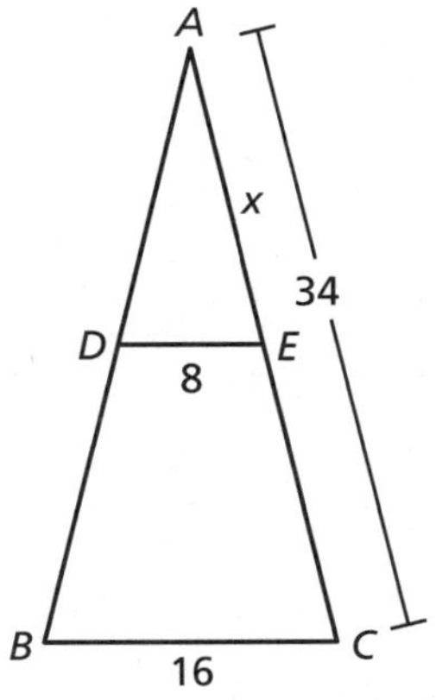

Copyright © Big Ideas Learning, LLC
All rights reserved.

Name ____ Date __________

15.3 Puzzle Time

Did You Hear About The...

A	B	C	D	E	F
G	H	I	J	K	L
M					

Complete each exercise. Find the answer in the answer column. Write the word under the answer in the box containing the exercise letter.

525 UP
15 HOLDUP
10% WIND
96 PANTS
22 SOCKS
25% WHEN
33 PAIR
200 HELD
37.5% IN
625 ROPE

Write and solve a proportion to answer the question.

A. What number is 25% of 60?

B. 30 is what percent of 80?

C. 20 is 40% of what number?

D. What number is 110% of 70?

E. 32 is what percent of 128?

F. What percent of 48 is 36?

G. 12 is what percent of 60?

H. 8% of what number is 16?

I. 42 is 8% of what number?

J. 30% of 140 is what number?

K. 150% of 22 is what number?

L. What number is 0.6% of 30?

M. The seventh-grade class needs to earn money for a trip to the amusement park. Of the 160 seventh-grade students, 60% participate in the fundraiser. How many students participate in the fundraiser?

20% CLOTHESPINS
50 THE
36% SUNSHINE
0.18 OF
120 CLOTHESLINE
75% TWO
100% RAINBOW
42 A
77 YARD
325 WRINKLES

Copyright © Big Ideas Learning, LLC
All rights reserved.

Activity 15.4 Start Thinking!

For use before Activity 15.4

Explain to a partner how to use the percent proportion to determine the number of points you need to get a score of 80% on a 45-point test.

Activity 15.4 Warm Up

For use before Activity 15.4

Write and solve a proportion to answer the question.

1. What percent of 15 is 6?
2. 22 is what percent of 40?
3. What number is 60% of 55?
4. 30% of 63 is what number?
5. 0.5% of what number is 0.425?
6. $\frac{3}{4}$ is 75% of what number?

Copyright © Big Ideas Learning, LLC
All rights reserved.

Lesson 15.4 Start Thinking!

For use before Lesson 15.4

Two students are working together to answer the question "what number is 5% of 30?".

One student says to use the percent proportion $\frac{30}{x} = \frac{5}{100}$ and the other student says to use the percent equation $a = 0.05(30)$.

Who is correct? Explain your reasoning. Find the number.

Lesson 15.4 Warm Up

For use before Lesson 15.4

Answer the question. Explain the method you chose.

1. What number is 30% of 60?

2. 25 is what percent of 75?

3. 42 is 105% of what number?

4. What number is 80% of 120?

5. 45 is what percent of 100?

6. 91 is 65% of what number?

Copyright © Big Ideas Learning, LLC
All rights reserved.

Name________________________________ Date__________

15.4 Practice A

Answer the question. Explain the method you chose.

1. 24 is what percent of 60?

2. 8 is 40% of what number?

Write and solve an equation to answer the question.

3. What number is 70% of 120?

4. 30 is what percent of 120?

5. 112 is 56% of what number?

6. 128 is what percent of 80?

7. What number is 140% of 45?

8. 15 is 6% of what number?

9. There are 35 competitors in a marathon. Sixty percent of these finished the race in under four hours. How many competitors finished the race in under four hours?

10. Your class is going on a field trip. Twenty-four students have turned in their permission slips so far. This is 80% of the students in the class. How many students are in the class?

11. You take a test with 32 questions on it. You answer 24 questions correctly. What percent of the questions do you answer correctly?

12. You have r rare coins, consisting of p pennies and n nickels.

    a. p is 20% of 190. How many pennies do you have?

    b. 190 is 200% of r. How many rare coins do you have?

    c. n is 60% of r. How many nickels do you have?

13. The table shows the sales receipt for your purchase.

    a. The items with a "T" next to the price are subject to sales tax. What percent sales tax did you pay?

    b. Calculate the price of the top.

    c. The price you paid for the top was 60% of the original price. What was the original price of the top?

Item	**Price**
top	p
earrings	\$ 3.00 T
socks	\$ 2.00
granola bar	\$ 0.50 T
Subtotal	\$13.00
Tax	\$ 0.21
Total	\$13.21

Tell whether the following statement is *true* or *false*. Explain your reasoning.

14. 120% of a whole number is always greater than the number.

15. You can find 0.5% of a number by multiplying the number by $\frac{5}{100}$.

Copyright © Big Ideas Learning, LLC
All rights reserved.

Name ______________________________ Date __________

15.4 Practice B

Answer the question. Explain the method you chose.

1. 27 is what percent of 90?

2. 7 is 5% of what number?

Write and solve an equation to answer the question.

3. 27 is 0.5% of what number?

4. What number is 125% of 240?

5. 1.4% of what number is 28?

6. 27 is what percent of 72?

7. During a given month, there was a total of 23.6 inches of rain. This was 250% of the average rainfall for that month. What is the average rainfall for that month?

8. To maintain an acceptable level of chlorine in your pool, you add 1.4 gallons of chlorine. This is 0.007% of the amount of water in your pool. How many gallons of water are in your pool?

9. You must attend a minimum of 85% of the practices in order to play in the playoffs. You have made 37 of the 42 practices. Will you be able to play in the playoffs?

10. You are in charge of the seventh grade graduation dinner. The table shows the results of a survey of students' meal preferences.

Choice	Percent
Chicken Nuggets	25%
Spaghetti	?
Pizza	45%
Fish Sticks	?

   a. 144 students chose pizza. How many students responded to the survey?

   b. How many students chose chicken nuggets?

   c. The number of students choosing fish sticks was 50% of the number of students choosing spaghetti. How many students chose fish sticks?

   d. How many students chose spaghetti?

11. What is 15% of 40% of $180?

12. There are 15 copies of a popular CD left to be sold in a store. This is between 1% and 1.5% of the original number of copies of the CD in the store. The original number of CDs was between what two numbers?

13. Tell whether the statement is *true* or *false*. Explain your reasoning.

If A is 45% of B, then the ratio $A : B$ is 9 : 20.

Find the percent to the nearest hundredth.

14. 16 is what percent of 38?

15. 50 is what percent of 38?

Copyright © Big Ideas Learning, LLC
All rights reserved.

Name____ Date__________

15.4 Enrichment and Extension

The Percent is Right: Closest Without Going Over

Preparation:

- Cut index cards to make 20 playing cards. Write each number from 1 to 20 on the cards.

To play:

- On your turn, choose a card and record that value. Your current score is the percent of the value chosen out of the total possible, which is 20.
- With each consecutive turn, add your new card value to your previous total. Also add 20 to the total possible. Then, calculate your new score.
- You may choose to draw as many as 4 times, replacing the cards each time. You may end your turn at any time. But, you may not go back and draw again after other players have gone.
- The player who comes closest to 50% without going over, wins. If players tie, the one who drew the fewest number of cards is the winner.

A sample score card is shown.

Turn	Card Value	Total Card Value	Total Possible	Score
1	7	7	20	35%
2	15	22	40	55%
3	6	28	60	46.6%

Complete the following exercises after playing a few rounds.

1. You draw a 5 on your first turn and a 17 on your second turn. Should you draw again? Explain your reasoning.
2. On your first 3 draws, you draw a 5, a 6, and a 10. Your opponent has already finished with a score of 42%. What is the lowest card you can draw and still win? What is the highest card you can draw and still win? What percent of the cards could you draw and still win the game?
3. What is the most you can get on your first three draws and still have a chance to win? Explain.
4. You are the first player to take a turn and you draw a 7. Should you draw again? Explain your reasoning. What if your first draw is an 8 or a 9?
5. If you have time, you can play multiple rounds with a different target percent score each round. Explain how a higher or lower target percent score would change the game.

Copyright © Big Ideas Learning, LLC
All rights reserved.

Name ______________________________ Date __________

15.4 Puzzle Time

Did You Hear About...

A	B	C	D	E	F
G	H	I	J	K	L
M	N				

Complete each exercise. Find the answer in the answer column. Write the word under the answer in the box containing the exercise letter.

42	HIS
150	THAT
52	TO
48	DOWN
15	THE
300	TEAM
75	MATCH
55	SET
145	ENDED
175	PLAYED
65	EAR
35	CHEER
93	FAN
82	TOO

Solve an equation to answer the question.

A. What percent of 60 is 9?

B. 12 is what percent of 30?

C. 62% of 150 is what number?

D. 8% of what number is 2?

E. What percent of 130 is 91?

F. 13% of 400 is what number?

G. 72 is what percent of 80?

H. 64% of what number is 48?

I. 175% of 8 is what number?

J. What percent of 200 is 290?

K. 156 is what percent of 60?

L. 0.5% of what number is 3?

M. Of the 60 students in the seventh grade, 70% own a pet. How many of the seventh grade students own a pet?

N. A pair of sunglasses is on sale for $26. The original price of the sunglasses was $40. What percent of the original price is the sale price?

17	ARE
25	WHO
225	THIS
260	UP
125	GAME
90	A
45	COACH
40	SPORTS
97	STARTED
70	LISTENED
20	RUNNING
600	BURNING
400	PLAYING
14	AND

Copyright © Big Ideas Learning, LLC
All rights reserved.

Activity 15.5 Start Thinking!

For use before Activity 15.5

Explain how a meteorologist uses percents.

Activity 15.5 Warm Up

For use before Activity 15.5

Write the decimal as a percent.

1. 0.45 **2.** 1.34 **3.** 0.549

4. 1.08 **5.** 0.985 **6.** 0.3225

Copyright © Big Ideas Learning, LLC
All rights reserved.

Lesson 15.5 Start Thinking!

For use before Lesson 15.5

Work with a partner to write and solve a real-life problem for the following situation: \$250 decreased by 35%.

Lesson 15.5 Warm Up

For use before Lesson 15.5

Find the new amount.

1. 15 inches increased by 20%

2. 145 gallons decreased by 60%

3. 70 meters increased by 80%

4. 150 grams decreased by 74%

5. 120 pounds decreased 5%

6. 40 liters increased by 25%

Copyright © Big Ideas Learning, LLC
All rights reserved.

Name____ Date__________

15.5 Practice A

Find the new amount.

1. 12 dogs decreased by 25%
2. 140 fluid ounces increased by 45%
3. 100 textbooks increased by 99%
4. 75 students decreased by 80%

Identify the percent of change as an *increase* or a *decrease*. Then find the percent of change. Round to the nearest tenth of a percent, if necessary.

5. 5 cups to 8 cups
6. 150 pounds to 135 pounds
7. 14 dollars to 10 dollars
8. 28 seconds to 23 seconds
9. $\frac{1}{3}$ to $\frac{2}{3}$
10. $\frac{1}{3}$ to $\frac{1}{6}$
11. Yesterday your bus ride to school took 10 minutes. Today your bus ride took 12 minutes. What is the percent of change?
12. Yesterday 270 concert tickets were sold. Today 216 tickets were sold.
    a. Find the percent of change in the number of tickets sold from yesterday to today.
    b. Use the percent of change from part (a) to predict the number of tickets sold tomorrow. Round to the nearest ticket, if necessary.
    c. Find the predicted percent of change in the number of tickets sold from yesterday to tomorrow. Round to the nearest tenth of a percent, if necessary.
13. This month a band has 6 musicians. This is a 50% increase from the number of musicians in the band last month. How many musicians were in the band last month?
14. The sides of a square garden are 8 feet long.
    a. You enlarge the garden to create a 25% increase in the length of each side. Find the new length of the sides.
    b. Find the percent of change in the perimeter of the garden. Round to the nearest tenth of a percent, if necessary.
    c. Find the percent of change in the area of the garden. Round to the nearest tenth of a percent, if necessary.

Copyright © Big Ideas Learning, LLC
All rights reserved.

Name ______________________________ Date __________

15.5 Practice B

Find the new amount.

1. 55 employees increased by 20%

2. 25° decreased by 60%

3. 15 customers increased by 200%

4. 4200 fans increased by 0.5%

Identify the percent of change as an *increase* or a *decrease*. Then find the percent of change. Round to the nearest tenth of a percent, if necessary.

5. 3.2 kilograms to 2.4 kilograms

6. 41 euros to 85 euros

7. $\frac{2}{7}$ to $\frac{4}{7}$

8. $\frac{5}{6}$ to $\frac{1}{3}$

9. Last month you swam the 50-meter freestyle in 28.38 seconds. Today you swam it in 27.33 seconds. What is your percent of change? Round to the nearest tenth of a percent, if necessary.

10. Last week 1200 burgers were served at the Burger Barn.

a. This week 1176 burgers were served. What is the percent of change?

b. Use the percent of change from part (a) to predict the number of burgers served next week. Round to the nearest whole number, if necessary.

11. The price of a share of a stock was $37.50 yesterday.

a. Today there was a price decrease of 4%. What is today's price?

b. Based on today's price in part (a), what percent of change is needed to bring the price back up to $37.50? Round to the nearest tenth of a percent, if necessary.

12. The table shows the membership of two scout troops.

Year	Troop A	Troop B
2010	14	21
2011	16	24

a. What is the percent of change in membership from 2010 to 2011 for Troop A? Round to the nearest tenth of a percent, if necessary.

b. What is the percent of change in membership from 2010 to 2011 for Troop B? Round to the nearest tenth of a percent, if necessary.

c. Which troop has the better record in terms of the number of new members?

d. Which troop has the better record in terms of the percent of change in membership?

Copyright © Big Ideas Learning, LLC
All rights reserved.

Name____ Date__________

15.5 Enrichment and Extension

Selling Super Slick Shoes

You work at Super Slick Shoes and are planning a new advertising campaign. You have been given information about the shoe sales of your company and your competitor. You want to be sure that you are making your company sound as good as possible.

1. Your athletic shoe department sold 135 thousand pairs of Super Sweet Sneakers this year, which is a 125% increase over last year's sales.

a. How many pairs of Super Sweet Sneakers did you sell last year?

b. By how many thousand pairs did the sales increase?

2. The athletic shoe department of your competitors sold 258 thousand pairs of Super Smooth Sneakers this year, which is a 50% increase over last year's sales.

a. How many pairs of Super Smooth Sneakers did they sell last year?

b. By how many thousand pairs did their sales increase?

3. For sneakers, would you use the *amount* of increase in sales or the *percent* of increase in sales to make your company sound better than your competitor? Explain your reasoning.

4. Your designer shoe department sold 175 thousand pairs of Super Sleek Sandals this year and anticipates a 70% increase in sales next year.

a. How many pairs of Super Sleek Sandals do you expect to sell next year?

b. By how many thousand pairs do you expect the sales to increase?

5. Your competitors sold 150 thousand pairs of Super Snazzy Sandals this year and anticipate an 80% increase in sales next year.

a. How many pairs of Super Snazzy Sandals do they expect to sell next year?

b. By how many thousand pairs do they expect their sales to increase?

6. For sandals, would you use the anticipated *amount* of increase in sales or the anticipated *percent* of increase in sales to make your company sound better than your competitor? Explain your reasoning.

Copyright © Big Ideas Learning, LLC
All rights reserved.

Name ______________________________ Date __________

15.5 Puzzle Time

Why Did the Picture Go To Jail?

Write the letter of each answer in the box containing the exercise number.

Find the new amount.

1. 42 customers increased by 50%
2. 70 pennies decreased by 20%
3. 30 inches increased by 30%
4. 125 pounds decreased by 4%
5. 15 acres decreased by 80%
6. 12 stamps increased by 125%
7. 68 days increased by 75%
8. 440 miles decreased by 95%

Identify the percent of change as an *increase* or a *decrease*. Then find the percent of change.

9. 20 ounces to 25 ounces
10. 10 hours to 6 hours
11. 15 feet to 33 feet
12. 80 books to 64 books
13. 64 minutes to 144 minutes
14. 16 skateboards to 2 skateboards
15. 55 videos to 77 videos
16. 13.6 kilometers to 23.8 kilometers

Answers for Exercises 1–8

T. 3	**O.** 119
S. 120	**D.** 27
F. 22	**E.** 56
M. 39	**Y.** 63

Answers for Exercises 9–16

O. 40% increase
B. 87.5% decrease
I. 20% decrease
R. 25% increase
E. 75% increase
M. 120% increase
D. 40% decrease
A. 125% increase

4	15	11	2	14	7	10	1		8	9	13	3	16	6		12	5

Copyright © Big Ideas Learning, LLC
All rights reserved.

Activity 15.6 Start Thinking!

For use before Activity 15.6

Why does a store mark up an item?

Why does a store then sometimes discount that item?

Activity 15.6 Warm Up

For use before Activity 15.6

Write and solve an equation to answer the question.

1. What is 40% of 238?
2. 28 is what percent of 70?
3. What is 34% of 240?
4. 5% of what number is 6?
5. What is 110% of 150?
6. 42 is 250% of what number?

Copyright © Big Ideas Learning, LLC
All rights reserved.

Lesson 15.6 Start Thinking!

For use before Lesson 15.6

You go to a store to buy a new pair of jeans. You find 2 pairs of jeans each on sale for a different price. Explain which is a better bargain.

Regular Price: $35; Discount: 30%

Regular Price: $40; Discount: 35%

Lesson 15.6 Warm Up

For use before Lesson 15.6

Find the cost to store, percent of markup, or selling price.

1. Cost to store: $20
Markup: 15%
Selling price: ___?___

2. Cost to store: $65
Markup: 30%
Selling price: ___?___

3. Cost to store: $100
Markup: 12%
Selling price: ___?___

4. Cost to store: $150
Markup: 5%
Selling price: ___?___

Copyright © Big Ideas Learning, LLC
All rights reserved.

Name______________________________ Date__________

15.6 Practice A

Copy and complete the table.

	Original Price	Percent of Discount	Sale Price
1.	\$75	30%	
2.	\$18	65%	
3.		30%	\$42
4.		55%	\$90
5.	\$35		\$28
6.	\$55		\$46.75

Find the cost to store or selling price.

7. Cost to store: \$65

Markup: 25%

Selling price: __?__

8. Cost to store: __?__

Markup: 80%

Selling price: \$122.40

9. The cost to a store for a box of cereal is \$2.50. The store is selling the box of cereal for \$3.50. What is the percent of markup?

10. A store pays \$120 for a bicycle.

a. The store has a 60% markup policy. What is the selling price of the bicycle?

b. The store is now going out of business and is selling all of the bicycles at a 30% discount. What is the sale price of the bicycle?

c. Will the store make money or lose money on the bicycle? How much?

11. The selling price of a skateboard is \$147. The store has a 75% markup policy. What is the cost of the skateboard to the store?

12. You buy a watch for \$60.

a. There is a 6% sales tax. What is your total cost for the watch?

b. Your friend buys the same watch a month later. It is now sold at a discount of 15%. What is the new sale price?

c. What is your friend's total cost for the watch including tax?

d. What is the percent of change in the total cost?

Copyright © Big Ideas Learning, LLC
All rights reserved.

Name ______________________________ Date __________

15.6 Practice B

Find the original price, discount, sale price, selling price, markup, or cost to store. Round to the nearest penny, if necessary.

1. Original price: \$130

Discount: 45%

Sale price: __?__

2. Original price: \$500

Discount: __?__

Sale price: \$175

3. Original price: __?__

Discount: 5%

Sale price: \$68.40

4. Cost to store: \$1600

Markup: 33%

Selling price: __?__

5. Cost to store: \$65

Markup: __?__

Selling price: \$91

6. Cost to store: __?__

Markup: 25%

Selling price: \$437.50

7. You are buying shoes online. The selling price is \$29.99. Round to the nearest penny, if necessary.

a. The sales tax is 6.5%. What is the total cost?

b. The cost of shipping is 15% of the total cost. What is the total cost plus shipping?

c. If the total cost plus shipping is greater than \$35, then you receive a 10% discount off the original selling price. Do you qualify? If so, what is the new total cost plus shipping?

8. You have a coupon for \$15 off a video game. You can use it on 2 separate days.

a. On Monday, the discounted price of your video game is \$22.99. What is the original price of the game?

b. What is the percent of discount to the nearest percent?

c. On Thursday, the discounted price of your video game is \$12.99. What is the original price of the game?

d. What is the percent of discount to the nearest percent?

9. You buy a bracelet for \$15. You sell it at a craft show for \$25. What is the percent of markup to the nearest percent?

Copyright © Big Ideas Learning, LLC
All rights reserved.

Name______________________________ Date__________

15.6 Enrichment and Extension

Would You Get the Job?

You are applying for a job selling electronics. The interviewer asks you to answer the following questions in order to prove that you would be able to do your job effectively.

1. A customer is purchasing last year's version of the Football Mania video game, which was originally \$55. It is on a sale rack that says 60% off. The customer has a coupon for 30% off his entire purchase. He incorrectly says, "I can't believe I'm getting this for 90% off!"

   a. Find the cost of the video game after the coupon.

   b What would be the cost of the video game at 90% off?

   c. How would you explain to the customer that he is not getting 90% off?

2. Your boss has decided to make room for new models of digital cameras by discounting old models. In order to sell them quickly, she wants to sell them at the store's original cost before markup. She asks you to change the price tags and make signs with the percent of discount. Knowing that the store's percent of markup on digital cameras is 40%, a co-worker suggests that you just mark them down by 40%.

   a. How would you explain to your co-worker that the store would actually lose money if you did this?

   b. The original price of a camera is \$175.50. What was the store's original cost on this camera?

   c. What percent off will you write on the sign in order to sell this camera at cost? Round to the nearest percent.

3. If a customer brings an ad that shows a competitor is selling an item for less, you are permitted to match the competitor's price as long as the percent of discount is 15% or less.

   a. The newest Blu-ray disc player is listed at \$307.99. You notice on your way to work that the same player is for sale at another store for \$298.79. Would you be allowed to match this store's price if asked? Explain.

   b. The store sells a wireless Internet router for \$74.75. A customer has printed out a page from the Internet that shows the same router for \$59.97 plus 5% shipping and handling. Can you match this price? Explain.

Copyright © Big Ideas Learning, LLC
All rights reserved.

Name ______________________________ Date __________

15.6 Puzzle Time

What Does A Monster Say When Introduced?

Write the letter of each answer in the box containing the exercise number.

Find the sale price or the selling price.

1. $99 watch with a 40% discount
2. $32 earrings marked up 80%
3. $59 cell phone with a 10% discount
4. $65 digital camera marked up 50%
5. $35 soccer cleats marked up 95%

Find the original price.

6. swimsuit discounted 65% on sale for $20.30
7. tennis racket marked up 50% to $36.75
8. backpack discounted 10% on sale for $37.80
9. tickets marked up 35% to $59.40
10. perfume discounted 25% on sale for $27

Find the percent of discount or markup.

11. $52 sweater on sale for $31.20
12. $23 rollerblades marked up to $41.40
13. $30 board game on sale for $20.10
14. $0.60 bottle of iced tea marked up to $1.35
15. $3.50 nail polish marked up to $8.75

Answers for Exercises 1–10

T. $68.25 **U.** $53.10
E. $97.50 **A.** $42
L. $44 **O.** $57.60
A. $58 **D.** $24.50
S. $36 **E.** $59.40

Answers for Exercises 11–15

P. 125% **Y.** 80%
E. 150% **T.** 33%
O. 40%

14	9	1	6	10	4	7		13	2		15	8	5		12	11	3

Copyright © Big Ideas Learning, LLC
All rights reserved.

Activity 15.7 Start Thinking!
For use before Activity 15.7

Discuss the pros and cons of having a credit card.

Activity 15.7 Warm Up
For use before Activity 15.7

Find the sale price for the item. Round to the nearest cent.

1. Original price: \$300; Discount: 25%
2. Original price: \$75; Discount: 40%
3. Original price: \$1300; Discount: 20%
4. Original price: \$95; Discount: 15%
5. Original price: \$725; Discount: 10%
6. Original price: \$845; Discount: 35%

Copyright © Big Ideas Learning, LLC
All rights reserved.

Lesson 15.7 Start Thinking!
For use before Lesson 15.7

You earned $150 babysitting. You want to open a savings account. What factors must you consider before opening an account?

Lesson 15.7 Warm Up
For use before Lesson 15.7

An account earns simple interest. (a) Find the interest earned. (b) Find the balance of the account.

1. $750 at 2% for 3 years
2. $300 at 6% for 2 years
3. $1400 at 4% for 5 years
4. $600 at 4.5% for 7 years
5. $550 at 8% for 6 months
6. $1200 at 3.5% for 6 months

Copyright © Big Ideas Learning, LLC
All rights reserved.

Name______________________________ Date__________

15.7 Practice A

An account earns simple interest. (a) Find the interest earned. (b) Find the balance of the account.

1. \$200 at 3% for 5 years

2. \$750 at 8% for 2 years

3. \$1600 at 5% for 1 year

4. \$500 at 12% for 6 months

Find the annual interest rate.

5. $I = \$18$, $P = \$150$, $t = 6$ years

6. $I = \$164.50$, $P = \$940$, $t = 2.5$ years

Find the amount of time.

7. $I = \$72$, $P = \$600$, $r = 4\%$

8. $I = \$174$, $P = \$1450$, $r = 8\%$

9. You deposit \$350 in a savings account. The account earns 2.5% simple interest per year. What is the balance after 2 years?

Find the amount paid for the loan.

10. \$1000 at 8% for 5 years

11. \$3500 at 10% for 2 years

12. You deposit \$2000 in a savings account earning 5% simple interest. How long will it take for the balance of the account to be \$3800?

13. Your parents charge a family ski trip of \$3000 on a credit card.

a. The simple interest rate is 20%. The charges are paid after 6 months. What is the amount of interest paid?

b. What is the total amount paid for the ski trip?

14. Your parents could have taken out a loan for the ski trip in Exercise 13.

a. The simple interest rate is 6% and the time for the loan is 2 years. What would have been the total amount paid for the \$3000 ski trip?

b. What would be the monthly payment, if there were equal monthly payments?

c. Which loan option costs less, the credit card or the loan?

15. You deposit \$1200 in an account earning 8% simple interest.

a. What is the account balance after 1 year?

b. At the end of the first year, you deposit the balance of the account in a CD (certificate of deposit) earning 8% simple interest. What is the account balance after another year?

Copyright © Big Ideas Learning, LLC
All rights reserved.

Name ______________________________ Date __________

15.7 Practice B

An account earns simple interest. (a) Find the interest earned. (b) Find the balance of the account.

1. $2600 at 3.2% for 4 years
2. $75,000 at 8.5% for 3 months

Find the annual interest rate.

3. $I = \$41.80, P = \$440, t = 2$ years
4. $I = \$893.75, P = \$5500, t = 30$ months

Find the amount of time.

5. $I = \$9.90, P = \$360, r = 5.5\%$
6. $I = \$2064, P = \$10,000, r = 6.88\%$

Find the amount paid for the loan.

7. $20,000 at 7.5% for 10 years

8. $6000 at 12% for 2.5 years

9. You deposit $2000 in an account. The account earns $120 simple interest in 8 months. What is the annual interest rate?

10. You put money in two different accounts for one year each. The total simple interest for the two accounts is $140. You earn 6% interest on the first account, in which you deposited $1000. You deposited $800 in the second account. What is the annual interest rate for the second account?

11. You deposit $1200 in an account.

    a. The account earns 2.7% simple interest rate. What is the balance of the account after 3 months?

    b. The interest rate changes, and your new balance now earns 2% simple interest rate. What is the balance of the account after the next 6 months? Round to the nearest penny, if necessary.

    c. The interest rate changes again, and your new balance now earns 2.6% simple interest rate. What is the balance of the account after an additional 3 months? Round to the nearest penny, if necessary.

    d. How much did the account earn in simple interest for the year?

    e. Based on the interest in part (d), what is the actual simple interest rate for the year? Round to the nearest tenth of a percent.

12. You purchase a new guitar and take out a loan for $450. You have 18 equal monthly payments of $28 each. What is the simple interest rate for the loan? Round to the nearest tenth of a percent, if necessary.

Copyright © Big Ideas Learning, LLC
All rights reserved.

Name______________________ Date__________

15.7 Enrichment and Extension

Buying the Car of Your Dreams

People who sell cars are usually good negotiators. So, being an educated consumer is important. Most car loans are based on *compound interest*, which means that you pay interest on your interest. You will be better able to negotiate a good deal on a car if you understand how compound interest is calculated.

Example: A loan for $15,560 is taken out for 5 years at a yearly interest rate of 7.2% that is compounded annually. (a) What is the balance after 5 years? (b) What is the monthly payment?

a. $B = P(1 + r)^t$ — B = balance, P = principal, r = interest rate (in decimal form), t = number of times the interest is compounded

$= 15{,}560(1 + 0.072)^5$ — Substitute 15,560 for P, 0.072 for r, and 5 for t.

$= 15{,}560(1.072)^5$ — Add.

$= 22{,}028.43$ — Simplify.

The balance after 5 years is $22,028.43.

b. There are 60 months in 5 years. So, divide the balance by 60.

$$\frac{22{,}028.43}{60} \approx 367.14$$

The monthly payment is $367.14.

1. A salesman offers two discount options on the car in the example. He can decrease the initial cost of the car by $500 or decrease the interest rate by 0.5%. Find the final cost of both options. Which is the better deal?

2. Find the final cost of the loan in the example if the interest were compounded monthly instead of annually. (*Hint:* Divide the yearly interest rate by 12 to find the monthly interest rate.) Why are some loans compounded monthly and even daily?

3. Find the price of your dream car on the Internet, in the newspaper, or from another source. Then research the current interest rate at a dealership or other lender and how often their interest is compounded. Calculate the final cost and monthly payment for your dream car on a 5-year loan.

Copyright © Big Ideas Learning, LLC
All rights reserved.

Name ______________________________ Date __________

15.7 Puzzle Time

What Question Can You Never Answer Yes To?

Write the letter of each answer in the box containing the exercise number.

Find the interest earned.

1. \$500 at 4% for 3 years

D. \$50 **E.** \$60 **F.** \$100

2. \$1200 at 7% for 5 years

A. \$420 **B.** \$240 **C.** \$400

3. \$750 at 6% for 18 months

T. \$58.50 **U.** \$67.50 **V.** \$56.75

4. \$1500 at 8.5% for 6 months

J. \$62.50 **K.** \$65.50 **L.** \$63.75

Find the annual interest rate.

5. $I = \$200, P = \$1000, t = 4$ years

E. 5% **F.** 6% **G.** 7%

6. $I = \$30, P = \$600, t = 2$ years

M. 1.5% **N.** 2.25% **O.** 2.5%

7. $I = \$150, P = \$2500, t = 9$ months

R. 6% **S.** 8% **T.** 9%

8. $I = \$75, P = \$800, t = 15$ months

D. 6.5% **E.** 7.5% **F.** 8.5%

Find the amount of time.

9. $I = \$144, P = \$400, r = 6\%$

P. 3 years **Q.** 4 years **R.** 6 years

10. $I = \$236.25, P = \$750, r = 4.5\%$

N. 3 years **O.** 5 years **P.** 7 years

11. $I = \$87.50, P = \$3500, r = 5\%$

W. 3 years **X.** 3 months **Y.** 6 months

12. $I = \$108.75, P = \$2000, r = 7.25\%$

A. 9 months **B.** 6 months **C.** 3 months

2	9	5		11	6	3		12	7	4	1	8	10

Copyright © Big Ideas Learning, LLC
All rights reserved.

Name______________________________ Date__________

Chapter 15

Technology Connection

For use after Section 15.4

Finding Percents with a Calculator

In Section 15.4, to solve the percent equation $a = p \bullet w$ you converted the percent p to fraction or decimal form. You could also use a calculator with a [%] key to solve the equation.

EXAMPLE What number is 55% of 176?

SOLUTION

Step 1 This problem can be modeled by the equation $a = 55\% \bullet 176$.

Step 2 On your calculator, enter 55 [%] [×] 176 [=].

ANSWER 96.8

EXAMPLE 42 is 15% of what number?

SOLUTION

Step 1 This problem can be modeled by the equation $42 = 15\% \bullet w$.

Step 2 On your calculator, enter 42 [÷] 15 [%] [=].

ANSWER 280

Use a calculator to solve the percent problem.

1. What number is 84% of 325?
2. What number is 250% of 4?
3. What number is 90% of 30?
4. What number is 15% of 450?
5. 75% of what number is 225?
6. 98 is 14% of what number?
7. 6.2 is 40% of what number?
8. 0.5% of what number is 2?
9. According to the NCAA, only 2% of high school athletes receive an athletic scholarship for college. Your school has a total of 250 athletes. How many athletes would you expect to receive a scholarship?
10. In a fundraiser, 15% of the money collected goes to your school. If your school earns $780, how much money is collected?

Copyright © Big Ideas Learning, LLC
All rights reserved.

Assessment Book

Name____ Date__________

Chapter 11 Quiz

For use after Section 11.3

Find the absolute value.

1. $|8|$

2. $|-3|$

Complete the statement using <, >, or =.

3. 4 ___ $|-8|$

4. $|-5|$ ___ $|-10|$

5. 6 ___ $|2|$

6. $|-3|$ ___ 3

Order the values from least to greatest.

7. $3, |-5|, -2, 1, -4$

8. $11, |-9|, 0, |8|, -5$

Evaluate the expression.

9. $-12 + 5$

10. $4 + (-2)$

11. $-3 + (-7)$

12. $-4 - 3$

13. $9 - (-2)$

14. $-3 - (-7)$

15. A scuba diver dives down 20 feet into the ocean. He then swims 11 feet back up towards the surface. What is the position of the scuba diver relative to the surface?

16. You and your friend play a video game. You have a final score of 40 points, and your friend has a final score of -21 points. By how many points did you win?

17. The table shows the score for a golfer on the first nine holes. What is his score through nine holes?

Hole	1	2	3	4	5	6	7	8	9
Score	−1	0	+2	+1	−2	−1	−1	+2	−1

Answers

1. ____________
2. ____________
3. See left.
4. See left.
5. See left.
6. See left.
7. ____________
8. ____________
9. ____________
10. ____________
11. ____________
12. ____________
13. ____________
14. ____________
15. ____________
16. ____________
17. ____________

Copyright © Big Ideas Learning, LLC
All rights reserved.

Name ______________________________ Date __________

Chapter 11 Quiz

For use after Section 11.5

Tell whether the value of the expression is *positive* or *negative* without evaluating.

1. $-36 \div 4$
2. $-5 \bullet 16$
3. $72 \div 8$
4. $-13(-9)$

Evaluate the expression.

5. $-8 \bullet 5$
6. $-3(-4)$
7. $9(-7)$
8. $-18 \div 6$
9. $\frac{48}{-12}$
10. $27 \div (-3)$

11. You lose 2 points every time you forget to write your name on a test. You have forgotten to write your name 4 times. What integer represents your change in points from forgetting to write your name?

12. The cost for a movie ticket is \$7. The cost is reduced to \$5 for groups of 10 or more. How many people must be in a group for them to save a total of \$30?

13. You are swimming each 25-yard length of a swimming pool 3 seconds faster than your personal best. What integer represents your change in time of your personal best after 200 yards?

14. Pete owes his brother \$35 from five weeks of borrowing money.

    a. What is the mean amount Pete borrowed from his brother each week?

    b. If Pete continues to borrow money, how much money will he owe after three more weeks?

15. Seven days ago, the gas tank on Rosa's car was filled with 20 gallons of gasoline. Today there are 6 gallons of gasoline.

    a. What integer represents the change of gasoline in the past seven days?

    b. What is the mean change in gasoline in gallons per day?

    c. What is the fuel efficiency, in miles per gallon, of Rosa's car if she drove 490 miles in the past seven days?

Answers

1. ____________
2. ____________
3. ____________
4. ____________
5. ____________
6. ____________
7. ____________
8. ____________
9. ____________
10. ____________
11. ____________
12. ____________
13. ____________
14. a. ____________
    b. ____________
15. a. ____________
    b. ____________
    c. ____________

Copyright © Big Ideas Learning, LLC
All rights reserved.

Name ____________________ Date ________

Chapter 11 Test A

Find the absolute value.

1. $|-1|$
2. $|15|$
3. $|-37|$

Complete the statement using $<$, $>$, or $=$.

4. $|-6|$ ___ 6
5. 0 ___ $|3|$
6. $|-5|$ ___ $|-9|$
7. One fish is 4 feet below sea level. Another fish is 3 feet below sea level. Write each position as an integer. Which integer is greater?

Evaluate the expression.

8. $6 + (-3)$
9. $8 + (-1) + (-3)$
10. $10 - (-3)$
11. $-9 - (-9)$

Multiply.

12. $7 \bullet (-4)$
13. $-2(-5)(-3)$

Divide, if possible.

14. $-12 \div (-4)$
15. $-18 \div 6$
16. $\dfrac{-16}{8}$
17. $0 \div (-10)$
18. You start hiking at an elevation that is 80 meters below base camp. You increase your elevation by 42 meters. What is the new elevation with respect to base camp?
19. The temperature falls from 3°C to -4°C. What is the difference in these temperatures?

Find the mean of the integers.

20. $5, -15, 7, -13, 1$
21. $-21, -34, -37, -52, -56$
22. $-3, -1, 0, 1, 3$

Answers

1. ____________
2. ____________
3. ____________
4. See left.
5. See left.
6. See left.
7. ____________

____
8. ____________
9. ____________
10. ____________
11. ____________
12. ____________
13. ____________
14. ____________
15. ____________
16. ____________
17. ____________
18. ____________
19. ____________
20. ____________
21. ____________
22. ____________

Copyright © Big Ideas Learning, LLC
All rights reserved.

Name ______________________________ Date __________

Chapter 11 Test A (continued)

Use mental math to solve the equation.

23. $m - 4 = 10$

24. $w - (-2) = 6$

25. $4 - c = -12$

26. $k - 3 = -9$

27. $n + (-8) = 5$

28. $8 - d = 14$

Find the next two numbers in the pattern.

29. 6, −12, 24, −48, …

30. −2, 20, −200, 2000, …

31. The table shows the temperature in Des Moines, Iowa, for certain times during a particular day.

Time	3 A.M.	8 A.M.	1 P.M.	5 P.M.	10 P.M.
Temperature	−15°F	−6°F	22°F	10°F	−11°F

a. What are the high and low temperatures?

b. Find the range of temperatures.

c. Find the change in temperature from 5 P.M. to 10 P.M.

d. Based on the given five temperatures, what is the average temperature for the day?

e. Explain why your answer to part (d) is not an accurate average temperature for the day.

Answers

23. __________

24. __________

25. __________

26. __________

27. __________

28. __________

29. __________

30. __________

31. a. __________

____

b. __________

c. __________

d. __________

e. See left.

Copyright © Big Ideas Learning, LLC
All rights reserved.

Name ______________________________ Date __________

Chapter 11 Test B

Find the absolute value.

1. $|12|$ 2. $|-30|$ 3. $|-44|$

Complete the statement using <, >, or =.

4. $|-3|$ ___ -7 5. 2 ___ $|-2|$ 6. 5 ___ $|-8|$

Order the values from least to greatest.

7. $|-3|, |-5|, -4, 7, |0|$ 8. $|-6|, 10, -2, -6, |-7|$

9. The temperature in St. Louis, Missouri, is 31°F. The temperature in Duluth, Minnesota, is –29°F. Is the temperature in St. Louis or Duluth closer to 0°F?

Evaluate the expression.

10. $(-124) + 25 + (-87)$ 11. $97 + (-118) + 0$

12. $-17 - (-56) + 14$ 13. $15 - (-98) + (-150)$

Multiply.

14. $-12 \bullet 5 \bullet (-3)$ 15. $-2(-21)(-5)$

Divide, if possible.

16. $-78 \div (-6)$ 17. $-65 \div 0$

18. The water level is 3 feet below your dock. The tide goes out, and the water level lowers 1 foot. A storm surge comes in, and the water level rises 2 feet. Write an integer to indicate the new water level.

19. Two integers, n and p, have a product of –24. What is the largest possible sum of n and p? Explain how you found your answer.

Evaluate the expression.

20. $-13 + (-56) \div 14 + 9$ 21. $-16 \bullet 3 - (-36) \div 12$

Answers

1. ____________
2. ____________
3. ____________
4. See left.
5. See left.
6. See left.
7. ____________
8. ____________
9. ____________
10. ____________
11. ____________
12. ____________
13. ____________
14. ____________
15. ____________
16. ____________
17. ____________
18. ____________
19. See left.
20. ____________
21. ____________

Copyright © Big Ideas Learning, LLC
All rights reserved.

Name ______________________________ Date __________

Chapter 11 Test B (continued)

Find the mean of the integers.

22. $-45, -20, -35, -25, -40$

23. $17, -25, 3, -29, -15, 7$

24. Your video game scores are 64, -13, 73, -5, and 36.

a. What were your lowest and highest scores?

b. What was your mean score?

25. What number must be added to 15, -6, and 12 to have a mean of 5? Explain.

Use mental math to solve the equation.

26. $-24 + m = -47$

27. $d - (-7) = -19$

Find the next two numbers in the pattern.

28. $4, -12, 36, -108, \ldots$

29. $-800, 400, -200, 100, \ldots$

30. The table shows a bank account balance for a 5-day period.

Day	Monday	Tuesday	Wednesday	Thursday	Friday
Balance	-21	-55	102	125	-16

a. Which day had the balance closest to zero?

b. Find the range of balances for the 5-day period.

c. What integer represents the change in balance from Monday to Tuesday?

d. Based on the given five balances, what is the average balance for the 5-day period?

31. Find values of a and b that make the statement $|a + b| = |a| + |b|$ false.

Answers

22. ____________

23. ____________

24. a. ____________

____

b. ____________

25. ____________

____

____

____

26. ____________

27. ____________

28. ____________

29. ____________

30. a. ____________

b. ____________

c. ____________

d. ____________

31. ____________

Copyright © Big Ideas Learning, LLC
All rights reserved.

Name______________________________ Date ________

Chapter 11 Standards Assessment

1. Two points are plotted on the number line below.

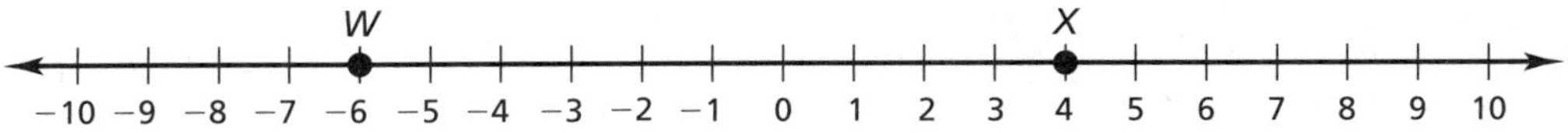

Which of the following has the **greatest** value?

A. $W - X$ **C.** $W + X$

B. $X - W$ **D.** $X \bullet W$

2. **GRIDDED RESPONSE** What is the value of the expression below?

$$12 - 6 \bullet 3$$

3. The temperature at 6:00 P.M. was 15°C. The temperature dropped 2°C per hour. What was the temperature at 4:00 A.M.?

F. −5°C **H.** 1°C

G. −1°C **I.** 5°C

4. Which list shows the integers in order from **least** to **greatest**?

A. −8, −5, 0, 2, 6 **C.** −5, −8, 0, 2, 6

B. 0, 2, −5, 6, −8 **D.** 0, −8, −5, 2, 6

5. The table shows the results of three plays in a football game. What is the net result of the three plays?

F. –16 yards

G. –8 yards

H. 8 yards

I. 12 yards

Football Game	
1st play	5 yards
2nd play	–9 yards
3rd play	12 yards

6. What is the value of the expression below?

$$|48 \div (-6)| + |-35 \div 7|$$

A. −13 **C.** 3

B. −3 **D.** 13

Copyright © Big Ideas Learning, LLC
All rights reserved.

Chapter 11 Standards Assessment (continued)

7. When $a = -9$ and $b = -6$, which expression has a value of -3?

F. $a + b$

G. $|a + b|$

H. $a - b$

I. $|a - b|$

8. Which of the following has the **greatest** value?

A. $|3|$

B. 0

C. $-|-5|$

D. $|-4|$

9. You record the daily low temperatures for four days at a ski lodge. What is the mean low temperature?

Low Temperature	
Thursday	–3°C
Friday	–6°C
Saturday	–5°C
Sunday	–2°C

F. –16°C

G. –4°C

H. 2.5°C

I. 4°C

10. **EXTENDED RESPONSE** In each of the equations or inequalities below, find all the **integer** values of x that make the equation or the pair of inequalities true. Explain your reasoning for each part.

Part A $|x| = 17$

Values of x ________________________

Part B $|x + 9| = 15$

Values of x ________________________

Part C $|x - 10| \leq 13$ and $|x - 10| \geq 9$

(Find the values of x that make **both** inequalities true.)

Values of x ________________________

Copyright © Big Ideas Learning, LLC
All rights reserved.

Chapter 11 Standards Assessment Item Analysis

1. **A.** The student thinks that the difference is 10 and is therefore greatest.
   B. Correct answer
   C. The student thinks that the sum must be positive and is therefore greatest.
   D. The student thinks that the product is 24 and is therefore greatest.

2. Correct answer: -6
   Common errors:
   - The student subtracts the smaller number from the larger product, getting an answer of 6.
   - The student subtracts before multiplying, getting an answer of 18.

3. **F.** Correct answer
   G. The student thinks that there are only 8 hours between 6:00 P.M. and 4:00 A.M. and finds $15 - 2(8)$.
   H. The student thinks that there are only 8 hours between 6:00 P.M. and 4:00 A.M. and thinks that $15 - 2(8) = 1$.
   I. The student thinks that $15 - 2(10) = 5$.

4. **A.** Correct answer
   B. The student disregards the signs and orders the numbers by magnitude only.
   C. The student thinks that -5 is less than -8.
   D. The student thinks that 0 is less than all other integers.

5. **F.** The student writes the sum as $5 - (9 + 12)$ which is -16.
   G. The student thinks that $5 + (-9) + 12 = -8$.
   H. Correct answer
   I. The student thinks the result of the 3rd play is the net result of all three.

6. **A.** The student thinks that quantities inside absolute value bars are always negative and finds the sum of -8 and -5.
   B. The student does not find the absolute value of the first expression and finds the sum of -8 and 5.
   C. The student thinks that $48 \div (-6) = 8$ because 48 is greater than 6. The student also ignores the absolute value bars and finds the sum of 8 and -5.
   D. Correct answer

7. **F.** The student thinks that $-9 + (-6) = -3$ because 9 is greater than 6 and 6 is being added to -9.
   G. The student thinks that $-9 + (-6) = -3$ because 9 is greater than 6 and 6 is being added to -9, then fails to apply the absolute value.
   H. Correct answer
   I. The student fails to apply the absolute value.

Copyright © Big Ideas Learning, LLC
All rights reserved.

8. A. The student ignores the absolute value bars in all options, so 3 is the only positive choice.

B. The student thinks that the presence of absolute value bars makes all quantities negative, so therefore 0 has the greatest value.

C. The student applies the absolute value to the negative sign that is outside the absolute value bars, getting 5 as the greatest value.

D. Correct answer

9. F. The student forgets to divide by 4.

G. Correct answer

H. The students subtracts the temperatures, $-3 - (-6) - (-5) - (2)$, and then divides by 4.

I. The student takes the absolute value of the numerator before dividing by 4.

10. 4 points The student demonstrates a thorough understanding of applying the concepts of absolute value to finding all integer values of variables that will make the equations or inequalities true. In Part A, the student correctly finds that 17 and -17 will make the equation true. In Part B, the student correctly finds that 6 and -24 will make the equation true. In Part C, the student correctly finds that $-3, -2, -1, 0, 1, 19, 20, 21, 22,$ and 23 will make the pair of inequalities true. The student provides clear and complete explanations of the reasoning used.

3 points The student demonstrates an understanding of applying the concepts of absolute value, but the student's work and explanations demonstrate an essential but less than thorough understanding. Students who correctly answer Parts A and B and provide a partially correct answer to Part C earn 3 points.

2 points The student demonstrates a partial understanding of applying the concepts of absolute value to finding all integer values of variables that will make the equations or inequalities true. The student's work and explanations demonstrate a lack of essential understanding. Students who correctly answer only two parts earn 2 points.

1 point The student demonstrates very limited understanding of applying the concepts of absolute value to finding all integer values of variables that will make the equations or inequalities true. The student's response is incomplete and flawed. Students who correctly answer only one part earn 1 point.

0 points The student provides no response, a completely incorrect or incomprehensible response, or a response that demonstrates insufficient understanding of applying the concepts of absolute value. Students whose answer to Part A is 17, to Part B is 6, and to Part C are 23 and 19 have demonstrated no understanding of absolute value and earn 0 points.

Copyright © Big Ideas Learning, LLC
All rights reserved.

Name____ Date__________

Chapter 11 Alternative Assessment

1. Your teacher has asked you to help Alec, a transfer student, who missed the teaching of absolute value and operations with integers. You agree to help him and you plan to work together at the library after school.

   a. Choose two sample problems to do with Alec to introduce him to the concept of absolute value of integers. Write a description of how you could use a number line to solve the sample absolute value problems.

   b. Choose four sample problems to do with Alec to introduce him to the concepts of addition, subtraction, multiplication, and division with integers.

   c. Write a description of how you could use counters to solve each of the sample integer operation problems in part (b).

   d. Write a description of how you could use a number line to solve each of the sample integer operation problems in part (b).

   e. Create a graphic organizer to help Alec remember everything you will teach him. Include the definition of the absolute value of an integer, and the sample problems used in part (a). Include the rules for operations with integers and the sample problems used in part (b).

Copyright © Big Ideas Learning, LLC
All rights reserved.

Name ______________________________ Date __________

Chapter 11

Alternative Assessment Rubric

Score	Conceptual Understanding	Mathematical Skills	Work Habits
4	Shows complete understanding of: • using a number line as a conceptual model for representing integers • operations with integers • absolute value of integers	Explains all integer operation concepts and rules correctly. Operational models represent given operations exactly.	Answers all parts of each problem. Work is neat and well organized.
3	Shows nearly complete understanding of: • using a number line as a conceptual model for representing integers • operations with integers • absolute value of integers	Explains most integer operation concepts and rules correctly. Operational models represent given operations adequately.	Answers all parts of each problem. Work is neat and easy to follow.
2	Shows some understanding of: • using a number line as a conceptual model for representing integers • operations with integers • absolute value of integers	Explains some integer operation concepts and rules correctly. Operational models do not represent given operations adequately.	Answers all parts of each problem. Work is sloppy and hard to follow.
1	Shows little understanding of: • using a number line as a conceptual model for representing integers • operations with integers • absolute value of integers	Does not explain integer operation concepts and rules correctly. Operational models do not represent given operations.	Does not answer all parts of each problem. Work is sloppy and hard to follow.

Copyright © Big Ideas Learning, LLC
All rights reserved.

Name ______________________________ Date __________

Chapter 12 Quiz

For use after Section 12.2

Write the rational number as a decimal.

1. $-\dfrac{9}{10}$

2. $-1\dfrac{3}{5}$

Write the decimal as a fraction or mixed number in simplest form.

3. -0.65

4. -3.75

Complete the statement using <, >, or =.

5. $\dfrac{14}{8}$ ___ $1\dfrac{3}{4}$

6. -5.65 ___ -5.6

7. $-6\dfrac{7}{8}$ ___ $-6.\overline{8}$

8. $-3\dfrac{11}{12}$ ___ $-3\dfrac{14}{15}$

Add. Write fractions in simplest form.

9. $-8\dfrac{3}{8} + 6\dfrac{1}{4}$

10. $-2\dfrac{3}{4} + \left(-1\dfrac{1}{3}\right)$

11. $-7.3 + (-3.6)$

12. $8.36 + (-4.825)$

13. The coldest temperature on record in Town A is $-3.33°$F.

The coldest temperature on record in Town B is $-3\dfrac{2}{5}°$F.

Which town has the colder temperature?

14. The table shows four transactions (in dollars) for a bank account. Positive number represent *deposits*, and negative numbers represent *withdrawals*. The balance before the transactions is \$75.50. What is the balance after the transactions?

Transactions	
Date	**Amount**
11/4	60.68
11/4	−25.16
11/7	−82.05
11/11	55.95

Answers

1. ____________
2. ____________
3. ____________
4. ____________
5. See left.
6. See left.
7. See left.
8. See left.
9. ____________
10. ____________
11. ____________
12. ____________
13. ____________
14. ____________

Copyright © Big Ideas Learning, LLC
All rights reserved.

Name ______________________________ Date __________

Chapter 12 Quiz

For use after Section 12.4

Subtract. Write fractions in simplest form.

1. $6.3 - 9.5$

2. $4.2 - (-2.6)$

3. $-5\frac{1}{2} - 2\frac{1}{3}$

4. $-3\frac{4}{9} - \left(-2\frac{1}{18}\right)$

Find the distance between the two numbers on the number line.

5.

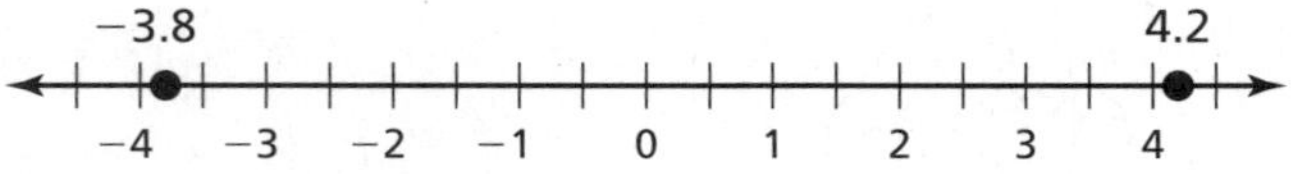

6.

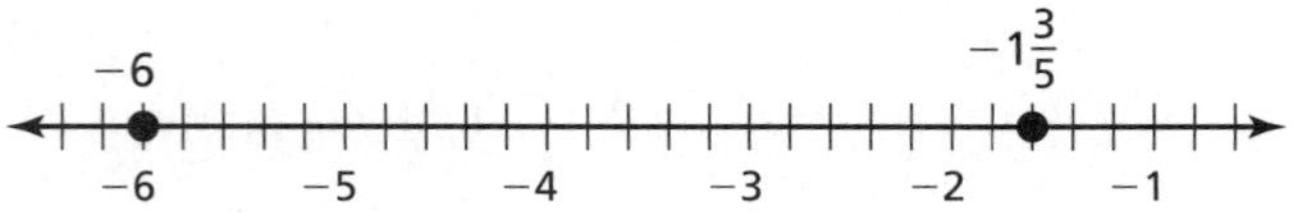

Multiply or divide. Write fractions in simplest form.

7. $4.5(-6.2)$

8. $-3\frac{3}{5} \div 2\frac{7}{10}$

9. $-4\frac{1}{6} \bullet \left(-3\frac{3}{5}\right)$

10. $-2.7 \div 0.9$

Evaluate.

11. $4.5 - 11.2 \div 2^3$

12. $-2\frac{3}{5} \times \frac{2}{3} - 1\frac{4}{5}$

13. A bottle contains 16.9 fluid ounces of water. After you take a sip, the bottle is $\frac{9}{10}$ full. How many fluid ounces is your sip?

14. Your class is holding a fundraiser for a local charity. After the first week, your class raised $\frac{1}{8}$ of your goal. After the second week, your class raised a total of $\frac{3}{5}$ of your goal. What fraction of the goal was raised during the second week of the fundraiser?

15. Stock A changed by $-\$5.70$ during the day. Stock B changed by $2\frac{1}{3}$ times that amount. How much did Stock B change during the day?

Answers

1. ____________
2. ____________
3. ____________
4. ____________
5. ____________
6. ____________
7. ____________
8. ____________
9. ____________
10. ____________
11. ____________
12. ____________
13. ____________
14. ____________
15. ____________

Copyright © Big Ideas Learning, LLC
All rights reserved.

Name________________________ Date__________

Chapter 12 Test A

Write the rational number as a decimal.

1. $-\frac{3}{11}$

2. $-3\frac{13}{20}$

Write the decimal as a fraction or a mixed number in simplest form.

3. 3.42

4. -0.35

Complete the statement using <, >, or =.

5. $-2\frac{7}{8}$ ___ $-2\frac{3}{4}$

6. -6.4 ___ -6.5

7. Your skateboard ramp is $2\frac{3}{8}$ feet high. Your friend's skateboard ramp is $2\frac{2}{5}$ feet high. Which skateboard ramp is higher?

Add or subtract. Write fractions in simplest form.

8. $5.73 - (-3.56)$

9. $-\frac{5}{3} + 2\frac{1}{3}$

10. $4\frac{2}{5} - \left(-6\frac{1}{2}\right)$

11. $-9.24 + (-2.89)$

Find the distance between the two numbers on the number line.

12.
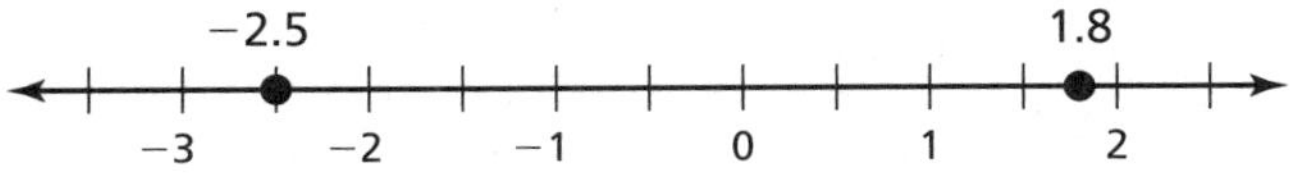

13.
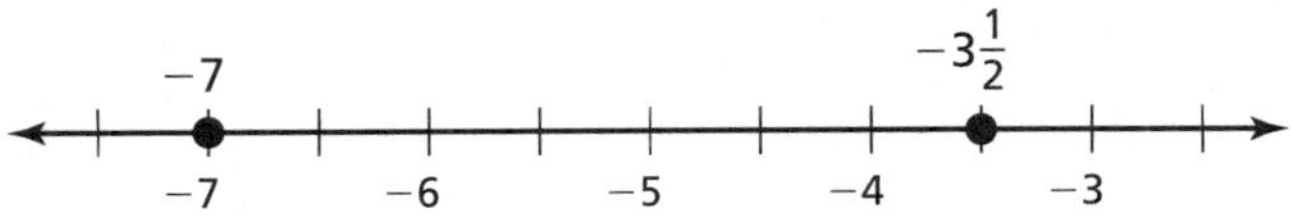

14. A gallon jug of milk is $\frac{3}{4}$ full. After breakfast the jug is $\frac{1}{12}$ full. Find the difference of the amounts before breakfast and after breakfast.

15. You buy a bag of dog food for \$12.59 and a bottle of dog shampoo for \$4.75. How much more did the dog food cost than the shampoo?

Answers

1. ____________
2. ____________
3. ____________
4. ____________
5. See left.
6. See left.
7. ____________
8. ____________
9. ____________
10. ____________
11. ____________
12. ____________
13. ____________
14. ____________
15. ____________

Copyright © Big Ideas Learning, LLC
All rights reserved.

Name ______________________________ Date __________

Chapter 12 Test A (continued)

Multiply or divide. Write fractions in simplest form.

Answers

16. $\left(-\frac{2}{5}\right)\left(-1\frac{1}{4}\right)$

17. $0.15 \times (-0.6)$

18. $-4.2 \div 12$

19. $-\frac{2}{7} \div \left(-\frac{8}{21}\right)$

20. $\frac{2}{3} \bullet \left(-2\frac{1}{4}\right) \bullet \left(-1\frac{1}{2}\right)$

21. $9.408 \div (-2.45)$

22. How many $\frac{2}{3}$-ounce packages of peanuts can be made with 8 ounces of peanuts? Explain how you found your answer.

23. Find the perimeter of the rectangle.

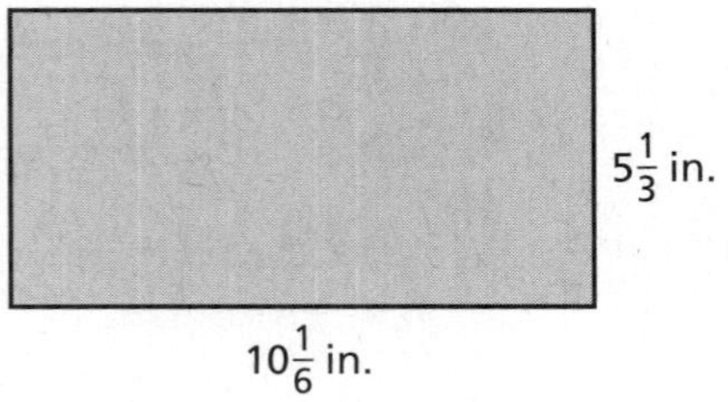

24. A 13.5-gallon gasoline tank is $\frac{4}{5}$ full. How many gallons will it take to fill the tank?

16. ____________
17. ____________
18. ____________
19. ____________
20. ____________
21. ____________
22. See left.
23. ____________
24. ____________

Copyright © Big Ideas Learning, LLC
All rights reserved.

Name______________________________ Date__________

Chapter 12 Test B

1. Order the numbers from least to greatest.

$$\frac{4}{5}, -1.6, -\frac{4}{3}, 0.9, \frac{2}{3}$$

Complete the statement using <, >, or =.

2. -2.34 ___ -2.43

3. $\frac{16}{11}$ ___ $1.\overline{45}$

Find the distance between the two numbers on the number line.

4.

5.

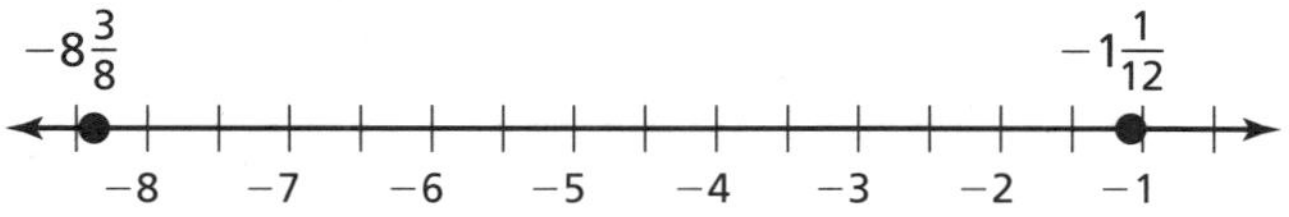

Add or subtract. Write fractions in simplest form.

6. $15.36 + (-12.095)$

7. $-7.91 - (-5.28)$

8. $-3\frac{7}{9} + \left(-2\frac{1}{3}\right)$

9. $\frac{13}{4} - \left(-4\frac{9}{10}\right)$

Evaluate the expression when $x = \frac{5}{8}$, and $y = -\frac{5}{3}$.

10. $x + y$

11. $y - x$

12. $-2x + y$

13. $3x + |y|$

14. You spend $3\frac{2}{3}$ hours hiking and an additional $\frac{3}{4}$ hour to rest.

a. How much time did you spend hiking and resting?

b. How much more time did you spend hiking than resting?

15. Find a repeating decimal between $-2\frac{1}{3}$ and $-2\frac{1}{5}$.

Answers

1. ______________
2. See left.
3. See left.
4. ______________
5. ______________
6. ______________
7. ______________
8. ______________
9. ______________
10. ______________
11. ______________
12. ______________
13. ______________
14. a. ______________
    b. ______________
15. ______________

Copyright © Big Ideas Learning, LLC
All rights reserved.

Name ______________________________ Date __________

Chapter 12

Test B (continued)

Answers

16. ________

17. ________

18. ________

19. ________

20. ________

21. ________

22. ________

23. See left.

24. ________

25. ________

Evaluate.

16. $3\frac{1}{5} - \left(-\frac{7}{2}\right) + (-1)$

17. $2.4 - |-3.61| - (-8.3)$

18. $-2\frac{1}{3} \times 5\frac{1}{4}$

19. $7.452 \div (-2.16)$

20. $\left(-\frac{3}{2}\right)^2 - \frac{1}{3}\left(3\frac{1}{2}\right)$

21. $0.1 \times (-10.5) - 4.76$

22. The table shows the changes in rainfall (in inches) from the monthly average of four months. What is the mean change?

Month	May	June	July	August
Change (inches)	1.05	−0.58	−2.12	−2.67

23. A recipe calls for $2\frac{1}{2}$ cups of sugar. You have $2\frac{1}{3}$ cups of sugar. Do you have enough sugar? If not, how much more sugar is needed? Explain your answer.

24. A 10.5-gallon aquarium is $\frac{2}{3}$ full. How many more gallons of water does it take to fill the aquarium?

25. How many 0.45-ounce packages of cinnamon can be made with 3.15 ounces of cinnamon?

Copyright © Big Ideas Learning, LLC
All rights reserved.

Name______________________________ Date__________

Chapter 12 Standards Assessment

1. What is the value of the following expression when $x = -9, y = -2$, and $z = 3$?

$$\frac{x - z}{-y}$$

A. -6 **C.** 3

B. -3 **D.** 6

2. Which of the following has the **greatest** value?

F. $5 - 9$ **H.** $-5 - 9$

G. $5 - (-9)$ **I.** $-5 - (-9)$

3. **GRIDDED RESPONSE** What is the next number in the pattern below?

$5, -10, 20, -40, \ldots$

4. Which of the following describes a correct method for finding the sum below?

$$-\frac{7}{2} + \frac{3}{4}$$

A. Write the sum of the numerators over the sum of the denominators. Simplify.

B. Rewrite using the LCD. Write the sum of the numerators over the sum of the denominators. Simplify.

C. Rewrite using the LCD. Write the sum of the numerators over the common denominator. Simplify.

D. Rewrite using the LCD. Write the difference of the numerators over the common denominator. Simplify.

5. What is the area, in square inches, of the rectangle?

F. $5\frac{13}{20}$

G. $6\frac{1}{10}$

H. $7\frac{13}{20}$

I. $11\frac{3}{10}$

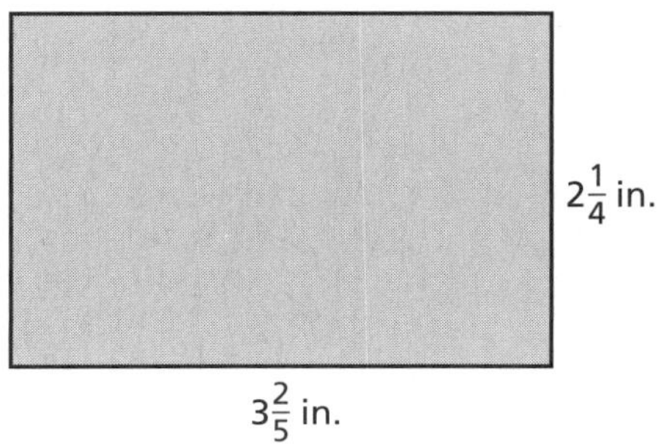

Copyright © Big Ideas Learning, LLC
All rights reserved.

Chapter 12 Standards Assessment (continued)

6. Which of the following has the **greatest** value?

A. $|4.4|$

B. $|-4.28|$

C. $-\left|\frac{9}{2}\right|$

D. $\left|-4\frac{3}{5}\right|$

7. A board is $6\frac{1}{2}$ feet long. You cut off a $1\frac{3}{4}$-foot piece and a $2\frac{2}{3}$-foot piece. How much of the original board is left?

F. $-2\frac{1}{2}$ feet

G. $-\frac{1}{6}$ foot

H. $2\frac{1}{12}$ feet

I. $4\frac{5}{12}$ feet

8. A truck driver drove 49 miles in $1\frac{1}{3}$ hours. What was the average speed, in miles per hour, for this trip?

A. $32\frac{2}{3}$

B. $36\frac{3}{4}$

C. $50\frac{1}{3}$

D. $65\frac{1}{3}$

9. **SHORT RESPONSE** When working overtime, an auto mechanic gets paid 1.5 times the normal hourly wage. Overtime is the amount of time worked beyond 40 hours in a week. The auto mechanic's normal hourly wage is \$21.00. Last week, the auto mechanic worked 46 hours.

Part A Write an expression to represent the amount of money earned last week. Explain your reasoning.

Expression ______________________________

Part B Evaluate the expression you wrote in Part A to find the amount of money earned last week. Show your work.

The auto mechanic earned ______ last week.

Copyright © Big Ideas Learning, LLC
All rights reserved.

Chapter 12 Standards Assessment Item Analysis

1. **A.** Correct answer
   B. The student incorrectly subtracts 3 from –9 and gets –6.
   C. The student incorrectly subtracts 3 from –9 and gets –6; then disregards the negative when substituting for y.
   D. The student disregards the negative when substituting for y.

2. **F.** The student thinks that $5 - 9 = 4$ and that all the other answer choices have negative values.
   G. Correct answer
   H. The student finds the least value of the choices.
   I. The student knows that answer choice I has a value of 4 but thinks that answer choice G has a value of -4.

3. Correct answer: 80

   Common error: The student thinks that the sign of each term alternates and the absolute value of each successive term increases by a power of 10. Because $|20| - |-10| = 10$ and $|-40| - |20| = 20,$ the next number is 70 because $|70| - |-40| = 30.$

4. **A.** The student just adds the numerators and the denominators.
   B. The student thinks that the denominator is the sum of the common denominators.
   C. Correct answer
   D. The student thinks that the negative addend means subtraction is used to find the numerator.

5. **F.** The student adds the dimensions instead of multiplying them.
   G. The student multiplies the whole parts, then multiplies the fractional parts, and then sums these results.
   H. Correct answer
   I. The student finds the perimeter.

6. **A.** The student thinks that $4\frac{3}{5} = 4.35$ and therefore thinks answer choice A is greater than answer choice D.
   B. The student disregards place value and thinks that 0.28 is greater than 0.4, so therefore answer choice B is greater than answer choice A. The student also disregards answer choice D.
   C. The student thinks that all absolute values are negative, so therefore answer choice C, which is the opposite of an absolute value, is the only positive answer choice.
   D. Correct answer

Copyright © Big Ideas Learning, LLC
All rights reserved.

Standards Assessment Item Analysis

(continued)

7. F. The student adds $1\frac{3}{4}$ and $2\frac{2}{3}$, then subtracts $6\frac{1}{2}$.

G. The student writes the expression using improper fractions, then subtracts the numerators and writes the result over the common denominator.

H. Correct answer

I. The student simply adds $1\frac{3}{4}$ and $2\frac{2}{3}$.

8. A. The student takes $\frac{1}{3}$ of 49 and subtracts this result from 49.

B. Correct answer

C. The student adds $1\frac{1}{3}$ to 49.

D. The student multiplies 49 by $1\frac{1}{3}$.

9. 2 points The student demonstrates a thorough understanding of writing and evaluating expressions using rational numbers. For Part A, the student correctly writes the expression $40(21) + 6(21)(1.5)$ and provides an appropriate explanation. For Part B, the student correctly gets a value of \$1029.00 and shows appropriate work.

1 point The student demonstrates a partial understanding of writing and evaluating expressions using rational numbers. The student writes a correct expression but does not evaluate it successfully, or the student does not write a correct expression but demonstrates the ability to evaluate an expression with rational numbers.

0 points The student demonstrates insufficient understanding of writing and evaluating expressions using rational numbers. The student does not write a correct expression and shows little or no evidence of being able to evaluate expressions.

Copyright © Big Ideas Learning, LLC
All rights reserved.

Name_______________________________ Date__________

Chapter 12 Alternative Assessment

1. Consider all the different rational numbers you studied in this chapter.

   a. Write three rational numbers as fractions from least to greatest. Remember that rational numbers can be negative.

   b. Use the numbers you just wrote in part (a) to write four expressions using addition or subtraction and four expressions using multiplication or division. Exchange lists with a partner and simplify each other's expressions. Describe a real-life situation that could be represented by one of the expressions.

   c. Repeat part (b) with three rational numbers written as decimals from least to greatest. For even more challenge, repeat part (b) again but use all six rational numbers to write expressions.

Copyright © Big Ideas Learning, LLC
All rights reserved.

Name ______________________________ Date __________

Chapter 12 Alternative Assessment Rubric

Score	Conceptual Understanding	Mathematical Skills	Work Habits
4	Shows complete understanding of: • writing and ordering rational numbers • operations with rational numbers	Writes and compares rational numbers accurately. Operates with rational numbers to simplify all expressions.	Answers all parts of the problem. Problem-solving steps are systematic. Work is very neat and well organized.
3	Shows nearly complete understanding of: • writing and ordering rational numbers • operations with rational numbers	Writes and compares rational numbers reasonably well. Operates with rational numbers to simplify most expressions.	Answers some parts of the problem. Problem-solving steps are somewhat systematic. Work is fairly neat and well organized.
2	Shows some understanding of: • writing and ordering rational numbers • operations with rational numbers	Is able to write or compare rational numbers, but not both. Operates with rational numbers to simplify some expressions.	Answers one part of the problem. Problem-solving steps are not very systematic. Work is not very neat or well organized.
1	Shows little understanding of: • writing and ordering rational numbers • operations with rational numbers	Is unable to write or compare rational numbers. Does not operate with rational numbers to simplify expressions.	Answers no parts of the problem. Problem-solving steps are not at all systematic. Work is not neat or well organized.

Copyright © Big Ideas Learning, LLC
All rights reserved.

Name____ Date____________

Chapter 13

Quiz

For use after Section 13.2

Identify the terms and like terms in the expression.

1. $10x + 5 + 3x + 1$

2. $-2n + 7n - r + 10r$

3. $-12h^2 - 4 + 9 - 3h^2$

4. $1.4c + 11.4 - 2c - 7.3c$

Simplify the expression.

5. $8v - 15v$

6. $7d + 5 - 4d$

7. $12x + 9 - 3x - 4$

8. $3(x - 4) + 5x$

Find the sum or difference.

9. $(3x - 5) + (-4x + 1)$

10. $6(-2.1k - 2) + (7k + 5)$

11. $(2m + 7) - (3 - 4m)$

12. $\frac{2}{3}(6c + 4) - (8c - 5)$

Factor out the coefficient of the variable.

13. $\frac{1}{2}d + 6$

14. $-3.6z - 10.8$

15. You and your friends order food from a menu where each item costs the same amount. Write an expression in simplest form that represents the total amount of money the order will cost.

16. Write an expression in simplest form for the area of the rectangle.

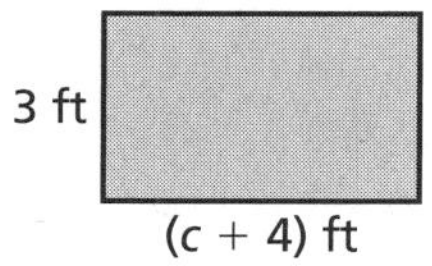

17. Eastside Bowling charges \$2.25 for shoes and \$3.00 per game. Westside Bowling charges \$1.75 for shoes and \$2.50 per game. Write an expression in simplest form that represents how much more Eastside Bowling charges than Westside Bowling.

Answers

1. ____________
____
2. ____________
____
3. ____________
____
4. ____________
____
5. ____________
6. ____________
7. ____________
8. ____________
9. ____________
10. ____________
11. ____________
12. ____________
13. ____________
14. ____________
15. ____________
16. ____________
17. ____________

Copyright © Big Ideas Learning, LLC
All rights reserved.

Name ____________________ Date ________

Chapter 13 Quiz

For use after Section 13.5

Solve the equation. Check your solution.

1. $b - 6 = -11$

2. $8 = q + 15$

3. $x + 4\frac{1}{3} = -2\frac{5}{6}$

4. $-2.5 + w = 3.7$

5. $\frac{n}{-5} = 7$

6. $-2p = \frac{4}{11}$

7. $17 = -7z + 3$

8. $\frac{a}{6} - \frac{1}{2} = \frac{1}{3}$

Write the word sentence as an equation. Then solve.

9. 3 less than a number h is -12.

10. The sum of a number b and 7.5 is 4.8.

11. The product of $-\frac{3}{4}$ and a number x is -27.

12. The quotient of a number m and -3.2 is 15.

13. The perimeter of the polygon is 27.4 centimeters. Write and solve an equation to find the unknown side length.

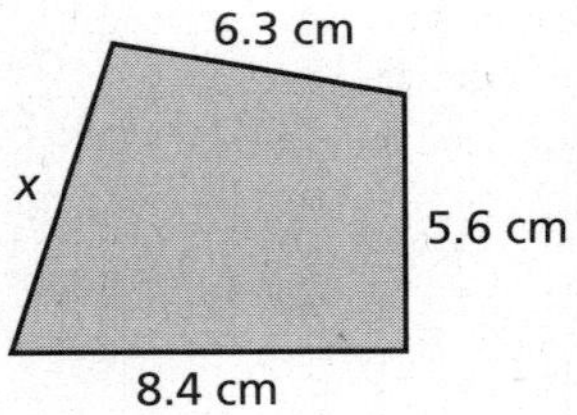

14. The bottom floor of a parking garage has an elevation of -36 feet. The top floor is 124 feet higher. What is the elevation of the top floor?

15. Store A sells jeans for $\frac{7}{8}$ of the price at Store B. Store A sells jeans for \$35. Write and solve an equation to find how much you save by buying jeans at Store A.

16. You are biking to your friend's house to do math homework. Halfway there, you realize you forgot your textbook, so you turn around. After biking 0.25 mile, you are one mile from your house. What is the distance between your house and your friend's house?

Answers

1. ______
2. ______
3. ______
4. ______
5. ______
6. ______
7. ______
8. ______
9. ______

   ______
10. ______

    ______
11. ______

    ______
12. ______

    ______
13. ______

    ______
14. ______
15. ______

    ______
16. ______

Copyright © Big Ideas Learning, LLC
All rights reserved.

Name____ Date__________

Chapter 13 Test A

Identify the terms and like terms in the expression.

Answers

1. $7j + 3 - 4j + 9$
2. $10a + 4b + 3a - 15b$
3. $x^2 - 3x + 4 + 2x^2 - x - 12$

Find the sum or difference.

4. $(4y + 3) - (y - 2)$
5. $(3p - 7) + (5p - 6)$
6. $(-2h + 1) + 2(3h - 4)$
7. $(5b - 9) - 3(8 - 2b)$

Factor out the coefficient of the variable.

8. $9x - 36$
9. $\frac{1}{5}k + 15$

10. Write an expression in simplest form that represents the perimeter of the polygon.

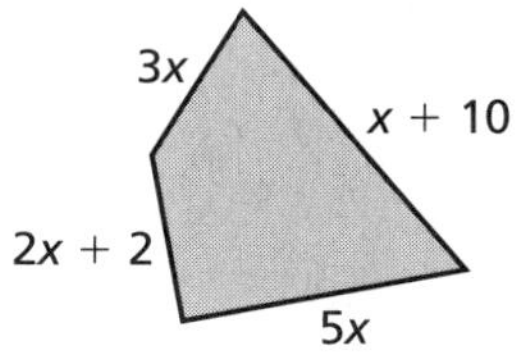

11. Write an expression in simplest form that represents the area of the picture frame.

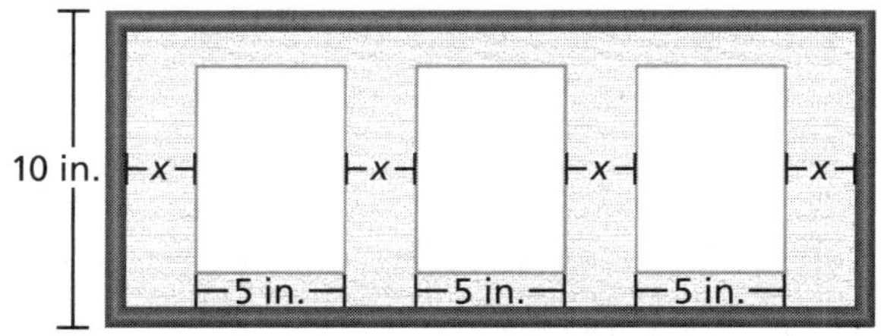

12. One page of a petition can hold p signatures. You were able to get $4p + 5$ people to sign the petition. Your friend was able to get $2p - 3$ people to sign the petition. Write an expression that represents the number of people that you and your friend got to sign the petition.

1. ____________
____
2. ____________
____
3. ____________
____
4. ____________
5. ____________
6. ____________
7. ____________
8. ____________
9. ____________
10. ____________
11. ____________
12. ____________

Copyright © Big Ideas Learning, LLC
All rights reserved.

Name ______________________________ Date __________

Chapter 13 Test A (continued)

Solve the equation. Check your solution.

13. $n - 6 = 21$

14. $-8.3 = d + 4.7$

15. $-0.5x = -4.3$

16. $-7 = \frac{z}{2} + 1$

17. $2y - 3y = 5$

18. $-2.9 = 3f + 4.3$

Write the word sentence as an equation. Then solve.

19. 11 more than a number q is negative 15.

20. The difference of a number m and 30 is 10.

21. One-third of a number t is equal to 7.

22. The quotient of 5 plus a number d and negative 2 is 14.

In Exercises 23–26, write an equation. Then solve.

23. The temperature is $-4°$F. A high pressure front increases the temperature to 8°F. By how many degrees did the temperature increase?

24. The monthly dues for a premium membership at a health club is $15 more than the cost of a standard membership. The premium membership is $40 per month. What is the cost of a standard membership?

25. A pack of cardinal flower seeds costs $4, and a pack of petunia flower seeds costs $2.50. You buy the same number of packs of each type of flower and spend $39. How many packs of each do you buy?

26. An egg carton holds 12 eggs. A breakfast buffet uses 96 eggs by 8 A.M. When the buffet ends at 10:30 A.M., a total of 156 eggs were used. How many cartons of eggs were used after 8 A.M.?

Answers

13. __________

14. __________

15. __________

16. __________

17. __________

18. __________

19. __________

____

20. __________

____

21. __________

____

22. __________

____

23. __________

____

24. __________

____

25. __________

____

26. __________

____

Copyright © Big Ideas Learning, LLC
All rights reserved.

Name______________________________ Date__________

Chapter 13 Test B

Identify the terms and like terms in the expression.

1. $m + 3n + 6m$

2. $-2t + 3t^2 - 10 - 14t$

3. $5x^2 + 7x - 3 + 10x^2 - 5x - 20$

Find the sum or difference.

4. $(3 - 7m) + 2(3.5m + 1)$

5. $\frac{1}{2}(8 - 6y) + \frac{1}{5}(10y - 25)$

6. $(4a - 5) - 8(-2.5a + 3)$

7. $\frac{2}{3}(7w + 4) - \frac{1}{3}(2w - 1)$

Factor out the coefficient of the variable.

8. $2.6x + 23.4$

9. $-\frac{3}{5}k - \frac{3}{10}$

10. Write an expression that represents the perimeter of the polygon.

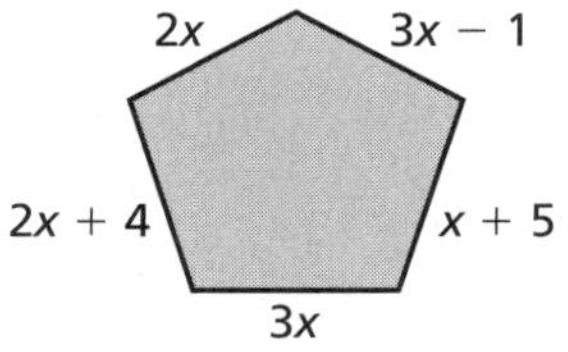

11. Write a formula for the area of the shaded region in terms of x.

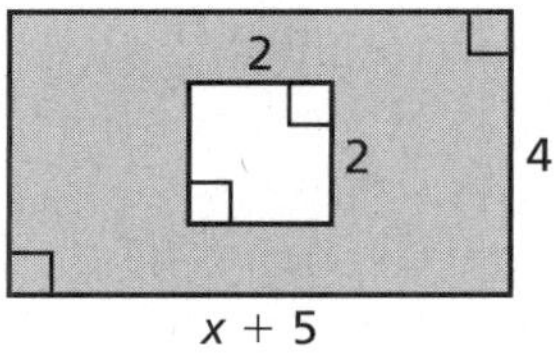

12. The expression $9n + 1$ represents the perimeter (in meters) of the triangle. Write an expression in simplest form that represents the measure of the third side.

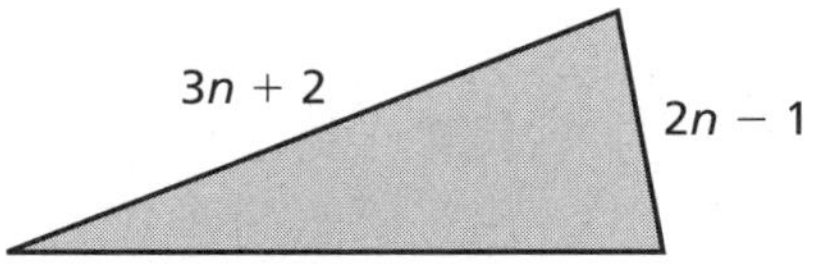

Answers

1. ____________

____

2. ____________

____

3. ____________

____

4. ____________

5. ____________

6. ____________

7. ____________

8. ____________

9. ____________

10. ____________

11. ____________

12. ____________

Copyright © Big Ideas Learning, LLC
All rights reserved.

Name ______________________________ Date __________

Chapter 13 Test B (continued)

Solve the equation. Check your solution.

13. $y + 14.6 = -31.75$

14. $m - \frac{5}{8} = -\frac{1}{4}$

15. $\frac{x}{6.2} = -2.1$

16. $-18v = -414$

Find the value(s) of *x*.

17. $5x - 8x = 15$

18. $3(x + 1) = -24$

19. $|x| + 3 = 14$

20. $|x + 2| = 7$

Write the word sentence as an equation. Then solve.

21. The sum of a number a and negative 12 is 6.

22. 45 equals the quotient of a number n and 3.

23. The difference of 2.1 and twice a number p is negative 4.7.

24. One-half the sum of a number z and $3\frac{1}{2}$ is $4\frac{1}{2}$.

25. You drink g 8-ounce glasses of water plus a 20-ounce bottle of water. You drank 76 ounces of water today. Write and solve an equation to determine the number of glasses of water you drank.

26. Your friend has owned a dog for 9 years. This is one year less than twice as long as he has owned a cat. How long has your friend owned a cat?

27. Litter cleanup volunteers form 4 groups containing v volunteers each. Then 10 more volunteers show up. The volunteers regroup into 5 groups each containing 6 people. How many volunteers were in each original group?

Answers

13. ____________

14. ____________

15. ____________

16. ____________

17. ____________

18. ____________

19. ____________

20. ____________

21. ____________

____

22. ____________

____

23. ____________

____

24. ____________

____

25. ____________

____

26. ____________

27. ____________

Copyright © Big Ideas Learning, LLC
All rights reserved.

Name______________________________ Date__________

Chapter 13

Standards Assessment

1. Which of the following equations is equivalent to the equation $-5(3x - 8) = -45$?

   A. $-15x - 40 = -45$

   B. $-15x + 40 = -45$

   C. $-15x - 8 = -45$

   D. $-15x + 8 = -45$

2. Which of the following lists is ordered from least to greatest?

   F. $11, |4|, |-2|, 0, -6$

   G. $0, |-2|, |4|, -6, 11$

   H. $-6, 0, |-2|, |4|, 11$

   I. $-6, |4|, |-2|, 0, 11$

3. **GRIDDED RESPONSE** What is the value of the expression below?

   $$-58.8 \div (-3.5 \times 2)$$

4. Which of the following describes a correct method for solving the equation below?

   $$-\frac{7}{2}n + 5 = -16$$

   A. Add 5 to both sides, then divide both sides by $\frac{2}{7}$.

   B. Subtract 5 from both sides, then multiply both sides by $-\frac{7}{2}$.

   C. Add –5 to both sides, then multiply both sides by $-\frac{2}{7}$.

   D. Subtract 5 from both sides, then add $\frac{7}{2}$ to both sides.

5. What is the area, in square inches, of a triangle with a base of $2\frac{1}{3}$ inches and a height of $2\frac{1}{4}$ inches?

   F. $2\frac{5}{8}$

   G. $4\frac{1}{24}$

   H. $4\frac{1}{12}$

   I. $5\frac{1}{4}$

Copyright © Big Ideas Learning, LLC
All rights reserved.

Name ______________________________ Date __________

Chapter 13 Standards Assessment (continued)

6. For which values of a and b is the following statement *not* true?

$$|a| - |b| = -5$$

A. $a = -3, b = 8$

B. $a = -2, b = -7$

C. $a = 3, b = 8$

D. $a = 7, b = 2$

7. A loon is flying at a height of 10 feet above a lake. The loon dives into the water to catch a fish that is at a depth of 6 feet. What is the change in elevation of the loon?

F. -16 feet

G. -10 feet

H. -4 feet

I. 16 feet

8. Which of the following is the sum of $\frac{1}{2}$ and $-\frac{2}{3}$?

A. $-1\frac{1}{6}$

B. $-\frac{1}{5}$

C. $-\frac{1}{6}$

D. $-\frac{1}{12}$

9. EXTENDED RESPONSE One year ago, Derek joined a health club. He paid a yearly membership fee of \$259.99 that covers all the club's services except use of the racquetball courts. Derek paid \$3.75 each time he used a racquetball court. For the entire year at the health club, Derek paid a total of \$436.24.

Part A Write an equation to represent the problem. Use r to represent the number of times Derek used one of the racquetball courts. Explain your reasoning.

Equation ______________________________

Part B Solve the equation you wrote in Part A to find how many times Derek used a racquetball court. Show your work.

Derek used a racquetball court ______ times.

Copyright © Big Ideas Learning, LLC
All rights reserved.

Chapter 13 Standards Assessment Item Analysis

1. **A.** The student incorrectly distributes the negative sign to the second term inside the parentheses, getting -40 instead of 40.
   B. Correct answer
   C. The student does not distribute the -5 to the second term inside the parentheses, leaving it as -8.
   D. The student distributes only the negative sign but not the 5 to the second term inside the parentheses, getting 8.
2. **F.** The student orders the list from greatest to least.
   G. The student ignores the signs and absolute value bars and orders by magnitude only.
   H. Correct answer
   I. The student thinks absolute values are negative.
3. Correct answer: 8.4
   Common error: The student ignores the parentheses and divides and then multiplies by 2. This gives an incorrect answer of 33.6.
4. **A.** The student adds 5 to both sides instead of subtracting, and then divides both sides by $\frac{2}{7}$ instead of $-\frac{7}{2}$.
   B. The student multiplies both sides by $-\frac{7}{2}$ instead of dividing.
   C. Correct answer
   D. The student thinks that $-\frac{7}{2}n$ means subtraction and that the required inverse operation is addition.
5. **F.** Correct answer
   G. The student multiplies the whole number parts and the fractional parts separately and finds the sum.
   $$\frac{1}{2}\left(2\frac{1}{3}\right)\left(2\frac{1}{4}\right) = (2 \bullet 2) + \left(\frac{1}{2} \bullet \frac{1}{3} \bullet \frac{1}{4}\right)$$
   H. The student forgets to multiply by $\frac{1}{2}$ and multiplies the whole number parts and the fractional parts separately and finds the sum.
   $$\left(2\frac{1}{3}\right)\left(2\frac{1}{4}\right) = (2 \bullet 2) + \left(\frac{1}{3} \bullet \frac{1}{4}\right)$$
   I. The student forgets to multiply by $\frac{1}{2}$.

Copyright © Big Ideas Learning, LLC
All rights reserved.

Standards Assessment Item Analysis
(continued)

6. A. The student ignores the absolute value bars and gets –11.

B. The student thinks absolute values are always negative and gets 5.

C. The student thinks absolute values are always negative and gets 5.

D. Correct answer

7. F. Correct answer

G. The student thinks the elevation change is only above water.

H. The student thinks the elevation change is $-10 - (-6) = -4$ feet.

I. The student thinks the elevation change is $10 - (-6) = 16$ feet.

8. A. The student finds the LCD but incorrectly adds the numerators.

$$\frac{1}{2} + \left(-\frac{2}{3}\right) = \frac{3}{6} + \frac{-4}{6} = \frac{-7}{6} = -1\frac{1}{6}$$

B. The student adds the numerators and denominators.

$$\frac{1}{2} + \left(-\frac{2}{3}\right) = \frac{1 + (-2)}{2 + 3} = \frac{-1}{5}$$

C. Correct answer

D. The student finds the LCD and then adds the numerators and denominators.

$$\frac{1}{2} + \left(-\frac{2}{3}\right) = \frac{3}{6} + \frac{-4}{6} = \frac{-1}{12}$$

9. 4 points The student demonstrates a thorough understanding of writing and solving two-step equations using decimal numbers. For Part A, the student correctly writes the equation $259.99 + 3.75r = 436.24$ and provides an appropriate explanation. For Part B, the student correctly gets a value of 47 for r, shows appropriate work, and states that Derek used a racquetball court 47 times.

3 points The student demonstrates an understanding of writing and solving two-step equations using decimal numbers, but the student's work and explanations demonstrate an essential but less than thorough understanding.

2 points The student demonstrates a partial understanding of writing and solving two-step equations using decimal numbers. The student's work and explanations demonstrate a lack of essential understanding.

1 point The student demonstrates a limited understanding of writing and solving two-step equations using decimal numbers. The student's response is incomplete and exhibits many flaws.

0 points The student provides no response, a complete incorrect or incomprehensible response, or a response that demonstrates insufficient understanding of writing and solving two-step equations using decimal numbers.

Copyright © Big Ideas Learning, LLC
All rights reserved.

Name____ Date__________

Chapter 13 Alternative Assessment

1. You and your friends eat at a local restaurant.

Menu	
Soda	$a
Water	$___
Soup	$___
Salad	$___
Hamburger	$___
Chicken	$___
Vegetable Pasta	$2b
Chocolate Cake	$b
Apple Pie	$c
Ice Cream	$(c-a)

a. Some of the prices are missing from the menu. You ask the waiter for the missing prices. He states the following:

- The cost of water is \$1 less than the cost of a soda.
- The cost of a bowl of soup is the difference of the cost of apple pie and \$1.
- The cost of a salad is the cost of a soda squared.
- The cost of a hamburger is the sum of the cost of chocolate cake and the cost of apple pie.
- The cost of chicken is twice the sum of the cost of chocolate cake and the cost of water.

Use the information to write an expression for each missing price on the menu. Fill in each expression on the line provided.

b. You order a soda, a salad, a hamburger, chocolate cake, and ice cream. Write and simplify an equation that represents the cost C of your meal.

c. You later find that $a = 2$, $b = 3$, and $c = 4$. Evaluate the equation from part (b) to find the cost C of your meal.

d. The formula $T = 1.2(C + 0.025C)$ gives the total amount of your bill including sales tax and tip. What is the total amount you spend on your meal?

2. When you solve a two-step equation, you must decide what you will do first. Usually one way of solving an equation may have fewer steps and less complex arithmetic than another. To explore this idea, consider the equation

$$3x + \frac{2}{3} = 7\frac{5}{6}.$$

a. Solve the equation by first subtracting $\frac{2}{3}$ from each side.

b. Solve the equation by first dividing each side by 3.

c. Compare and contrast both ways of solving the equation. Use your analysis to describe whether subtracting first or dividing first is the more efficient process to use for solving an equation like this one.

Copyright © Big Ideas Learning, LLC
All rights reserved.

Name ______________________ Date ________

Chapter 13 Alternative Assessment Rubric

Score	Conceptual Understanding	Mathematical Skills	Work Habits
4	Shows complete understanding of: • writing, simplifying, and evaluating algebraic expressions and equations • solving two-step equations	Writes, simplifies, and evaluates all expressions and equations correctly. Does compare/contrast analysis completely and accurately.	Answers all parts of each problem. Problem-solving steps are systematic. Work is very neat and well organized.
3	Shows nearly complete understanding of: • writing, simplifying, and evaluating algebraic expressions and equations • solving two-step equations	Writes, simplifies, and evaluates most expressions and equations correctly. Does compare/contrast analysis reasonably well.	Answers some parts of each problem. Problem-solving steps are somewhat systematic. Work is fairly neat and well organized.
2	Shows some understanding of: • writing, simplifying, and evaluating algebraic expressions and equations • solving two-step equations	Writes, simplifies, and evaluates some expressions and equations correctly. Is able to compare or contrast methods, but not both.	Answers parts of one problem. Problem-solving steps are not very systematic. Work is not very neat or well organized.
1	Shows little understanding of: • writing, simplifying, and evaluating algebraic expressions and equations • solving two-step equations	Does not write, simplify, or evaluate expressions or equations correctly. Does not compare or contrast any part of the equation-solving process.	Answers no parts of either problem. Problem-solving steps are not at all systematic. Work is sloppy and hard to follow.

Copyright © Big Ideas Learning, LLC
All rights reserved.

Name____________________ Date__________

Chapter 14 **Quiz**

For use after Section 14.3

Write the ratio as a fraction in simplest form.

1. 24 messages : 10 messages
2. 5 meters to 20 meters

Use the ratio table to find the unit rate with the specified units.

3. miles per gallon

Gallons	0	2	4	6
Miles	0	31	62	93

4. cost per box

Boxes	3	6	9
Cost	\$3.60	\$7.20	\$10.80

Tell whether the ratios form a proportion.

5. $\frac{4}{7}, \frac{24}{35}$
6. $\frac{11}{12}, \frac{33}{36}$

Tell whether the two rates form a proportion.

7. 25 cars in 5 days; 60 cars in 12 days
8. 14 books in 2 boxes; 20 books in 3 boxes

Use the table to write a proportion.

9.

	Cashews	Peanuts
Dollars	12	16
Pounds	3	p

10.

	Monday	Tuesday
Emails	e	30
Hours	8	10

Solve the proportion.

11. $\frac{x}{10} = \frac{4}{5}$
12. $\frac{8}{9} = \frac{p}{81}$

13. The number of pictures your printer can print are shown in the table. Find the rate in pictures per minute.

Minutes	2	4	6	8
Pictures	16	32	48	64

14. On Monday, you swim 12 laps in 30 minutes. On Tuesday, you swim 15 laps in 45 minutes. Are these rates proportional? Explain.

15. A chemical compound requires 8 ounces of Chemical A and 12 ounces of Chemical B. A mixture contains 24 ounces of Chemical A and 30 ounces of Chemical B. How can you fix the mixture to make the chemical compound?

16. In an animal shelter, the ratio of dogs to cats is 5 to 3. There are 25 dogs. Write and solve a proportion to find the number c of cats.

Answers

1. ________
2. ________
3. ________
4. ________
5. ________
6. ________
7. ________
8. ________
9. ________
10. ________
11. ________
12. ________
13. ________
14. ________

____

____

____

15. ________

____

16. ________

____

Copyright © Big Ideas Learning, LLC
All rights reserved.

Name ______________________ Date __________

Chapter 14 Quiz

For use after Section 14.6

Solve the proportion.

1. $\frac{7}{3} = \frac{b}{18}$ 2. $\frac{10}{9} = \frac{5}{k}$ 3. $\frac{3.6}{m} = \frac{1.2}{3.6}$

Find the slope of the line.

4.

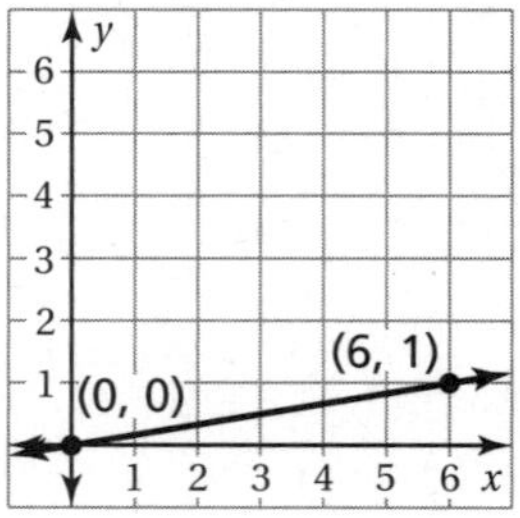

5.

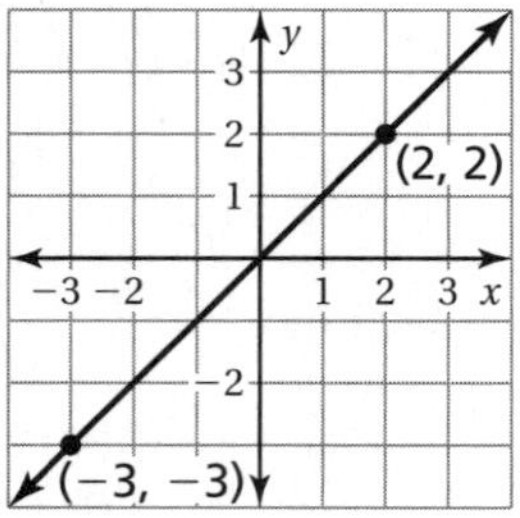

Graph the data. Then find and interpret the slope of the line through the points.

6.

Day, *x*	0	2	4	6
Inches, *y*	0	8	16	24

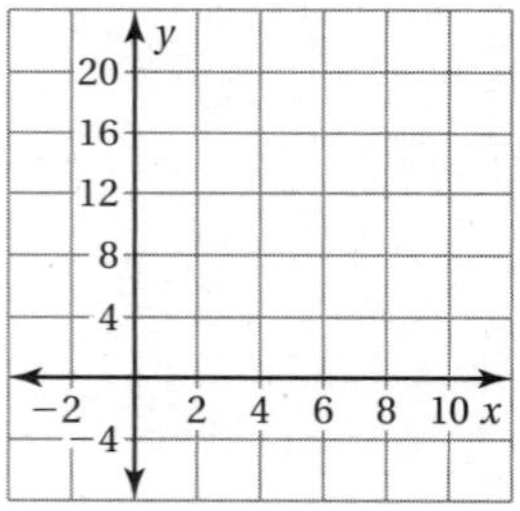

7.

Pounds, *x*	3	4	5	6
Cost, *y*	6	8	10	12

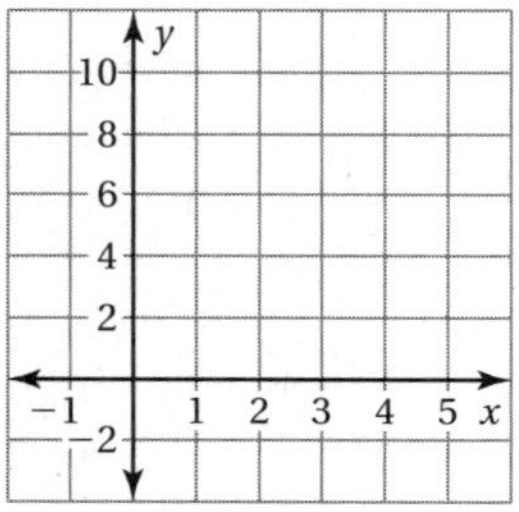

Tell whether *x* and *y* show direct variation. Explain your reasoning.

8. $xy = 5$ 9. $2x = 3y$

10. There are 25 students in a classroom. The ratio of girls to boys is 3 to 2. How many boys and how many girls are there in the classroom?

11. The graph shows the distance you and your friend drive on a trip.

    a. Find and interpret the slope of each line.

    b. How much faster are you traveling than your friend?

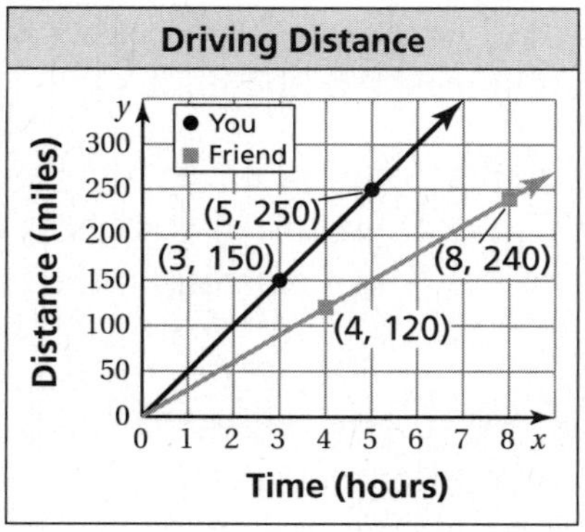

Answers

1. __________
2. __________
3. __________
4. __________
5. __________
6. See left.
7. See left.
8. __________
9. __________
10. __________
11. a. __________
    b. __________

Copyright © Big Ideas Learning, LLC
All rights reserved.

Name ______________________________ Date __________

Chapter 14 Test A

Write the ratio as a fraction in simplest form.

1. 15 girls to 6 boys
2. 24 players : 3 teams

Find the unit rate.

3. 405 rotations in 5 minutes
4. 72 ounces for 12 servings

Tell whether the ratios form a proportion.

5. $\frac{8}{24}, \frac{5}{15}$
6. $\frac{3}{7}, \frac{12}{21}$

7. The table shows the rates for sending text messages. Are the rates proportional? Explain.

Messages	Cost
50	$6
75	$9
100	$12

Use a graph to tell whether x and y are in a proportional relationship.

8.

x	2	4	6	8
y	1	2	3	4

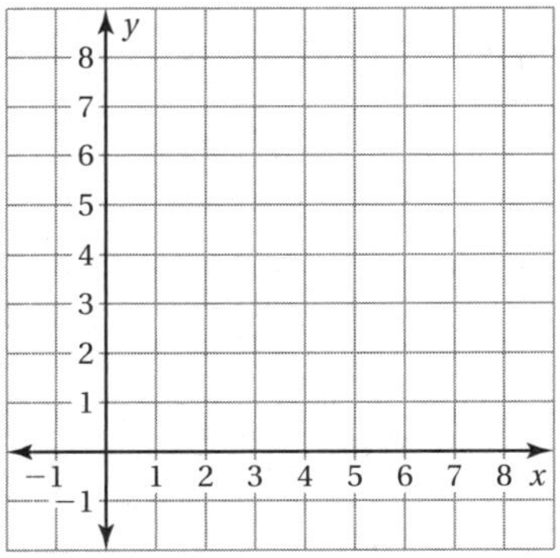

9.

x	3	6	9	12
y	7	13	19	25

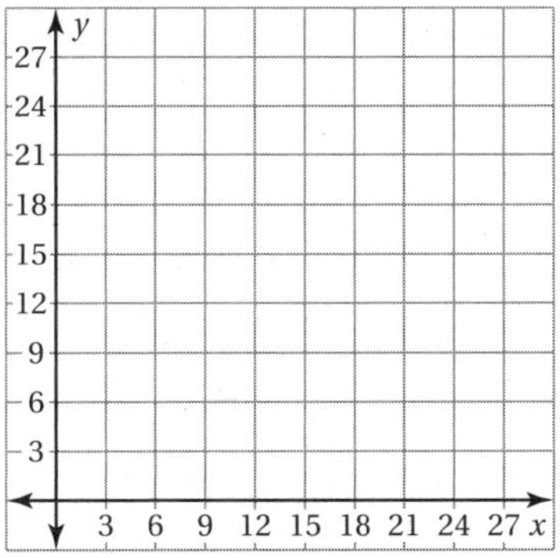

Use the table to write a proportion.

10.

	Serving A	Serving B
Servings	2	5
Calories	310	c

11.

	Saturday	Sunday
Minutes	m	32
Miles	3	4

12. You can buy 5 stickers for $3. Write a proportion that gives the cost c if you buy 12 stickers.

Answers

1. ____________
2. ____________
3. ____________
4. ____________
5. ____________
6. ____________
7. ____________
____
____
8. See left.
____
9. See left.
____
10. ____________
11. ____________
12. ____________

Copyright © Big Ideas Learning, LLC
All rights reserved.

Name ______________________________ Date __________

Chapter 14 Test A (continued)

Solve the proportion.

13. $\frac{2}{3} = \frac{n}{12}$ **14.** $\frac{33}{p} = \frac{3}{28}$ **15.** $\frac{k}{6} = \frac{15}{18}$ **16.** $\frac{2}{3} = \frac{3}{q}$

Graph the line that passes through the two points. Then find the slope of the line.

17. (0, 0), (3, 2)

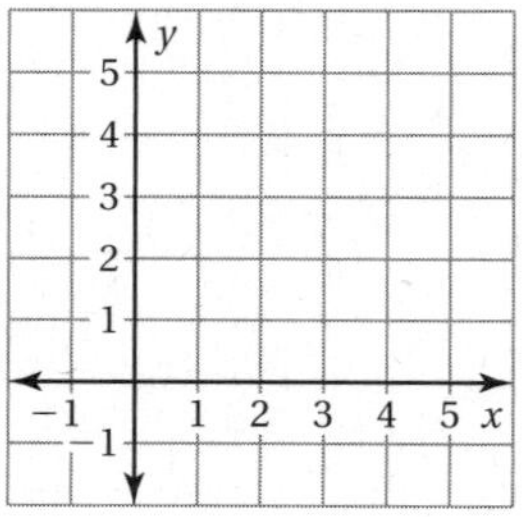

18. (4, 4), (8, 8)

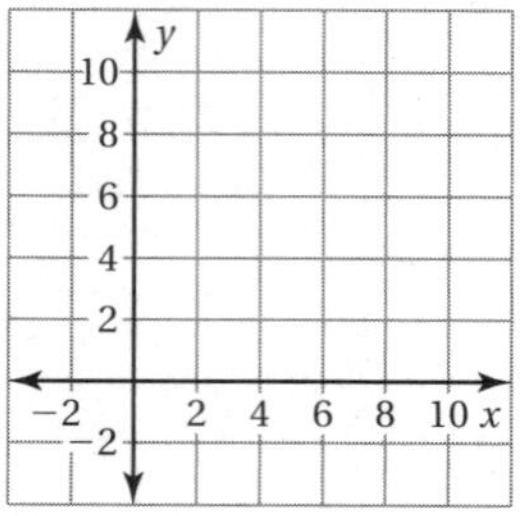

Tell whether *x* and *y* show direct variation. Explain your reasoning.

19.

x	1	2	3	4
y	4	8	12	16

20. $y + x = 1$

21. $\frac{y}{7} = \frac{2}{x}$

22. Your baseball team has won 6 games and lost 4 games. If the team does not lose any more games, how many games must the team win to have a win : loss ratio of 2 : 1? Explain your answer.

23. It costs $145 for 10 people to attend a concert. How much does it cost a group of 8 people?

24. The graph shows the calories burned for hiking and downhill skiing.

a. Find the slope of each line.

b. How many more calories do you burn per minute downhill skiing than hiking?

c. How many calories would you burn if you went downhill skiing for 80 minutes?

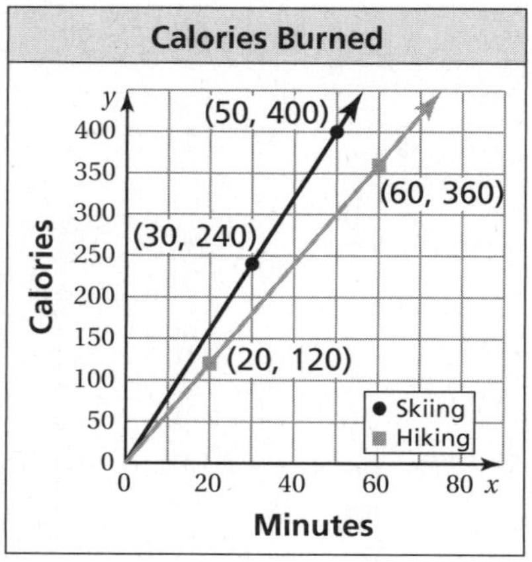

Answers

13. ____________

14. ____________

15. ____________

16. ____________

17. See left.

____

18. See left.

____

19. ____________

____

20. ____________

____

21. ____________

____

22. See left.

23. ____________

24. a. ____________

____

b. ____________

c. ____________

Copyright © Big Ideas Learning, LLC
All rights reserved.

Name____ Date__________

Chapter 14 Test B

Write the ratio as a fraction in simplest form.

1. 48 worksheets : 12 students
2. 35 frogs to 21 lizards

Find the unit rate.

3. 240 kilometers in 2.5 hours
4. $15 for 4 quarts

Tell whether the ratios form a proportion.

5. $\frac{56}{20}, \frac{24}{10}$
6. $\frac{5}{8}, \frac{42.5}{68}$

Tell whether *x* and *y* are proportional.

7.

x	5	10	15	20
y	1	3	5	7

8.

x	2	3	4	5
y	5	7.5	10	12.5

9. The table shows the different rates to ship books through the mail. Are the rates proportional? Explain.

Pounds	Cost
4	$3.55
6	$4.33
8	$5.11

Use the table to write a proportion.

10.

	512 MB MP3 Player	2 GB MP3 Player
Hours	17	68
Songs	*s*	1000

11.

	Tank A	Tank B
Fish	6	*f*
Gallons	10	55

12. You can buy 5 pounds of grapes for $9.95. Write a proportion that gives the cost *c* if you buy 4 pounds of grapes.

Solve the proportion.

13. $\frac{w}{84} = \frac{5}{7}$
14. $\frac{2.3}{1.8} = \frac{a}{18}$
15. $\frac{8}{25} = \frac{3}{d}$
16. $\frac{t}{6.5} = \frac{1.2}{1.3}$
17. $\frac{3x}{7} = \frac{8}{21}$
18. $\frac{16}{10} = \frac{n + 2}{5}$

Answers

1. ____________
2. ____________
3. ____________
4. ____________
5. ____________
6. ____________
7. ____________
8. ____________
9. ____________
____
____
____
10. ____________
11. ____________
12. ____________
13. ____________
14. ____________
15. ____________
16. ____________
17. ____________
18. ____________

Copyright © Big Ideas Learning, LLC
All rights reserved.

Name ____________________ Date __________

Chapter 14 Test B (continued)

Graph the line that passes through the two points. Then find the slope of the line.

19. $(-4, -3), (4, 3)$

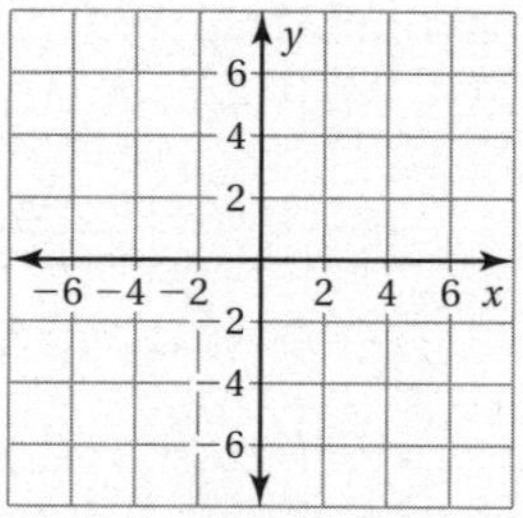

20. $(2, 1), (6, 3)$

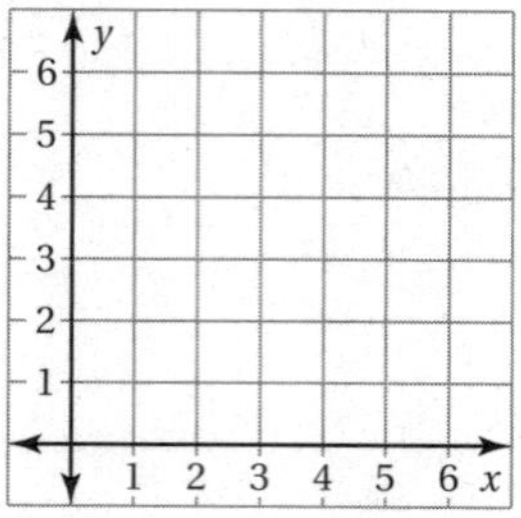

Tell whether *x* and *y* show direct variation. Explain your reasoning.

21. $5x - 3y = 0$

22. $x = \dfrac{y - 2}{9}$

The variables *x* and *y* vary directly. Use the values to find the constant of proportionality and write an equation that relates *x* and *y*.

23. $y = 4; x = 6$

24. $y = 2; x = 10$

25. Write a direct variation equation that relates x miles to y kilometers.

26. As part of a pancake recipe, you mix $\frac{3}{4}$ cup of milk for every 1 cup of flour to make 7 cups of batter. How much of each ingredient do you use?

27. You earn \$102 for doing 12 hours of yard work. Your friend earns \$120 working at a store for 15 hours.

a. Who has a greater hourly rate of pay?

b. What would you earn if you did 15 hours of yard work and were paid at your same hourly rate?

28. A line has a slope of 5. It passes through the points $(1, 4)$ and $(6, y)$. What is the value of y? Explain how you found your answer.

29. There are 25 computers in a math lab. There are 6 activity booklets for every 2 computers in the math lab. How many activity booklets are in the math lab?

Answers

19. See left.
____
20. See left.
____
21. ____________
____
22. ____________
____
23. ____________
____
24. ____________
____
25. ____________
26. ____________
____
27. a. ____________
b. ____________
28. See left.
29. ____________

Copyright © Big Ideas Learning, LLC
All rights reserved.

Name________________________________ Date__________

Chapter 14

Standards Assessment

1. Suppose that the equation below is true.

$$\frac{a}{b} = 5$$

Which of the following equations is also true?

A. $\frac{a + 2}{b + 2} = 5$

B. $\frac{a - 2}{b - 2} = 5$

C. $\frac{2a}{2b} = 5$

D. $\frac{2a}{2b} = 2(5)$

2. Which of the following has the **greatest** value?

F. $9 + (-9)$

G. $9 - (-9)$

H. $9 \bullet (-9)$

I. $9 \div (-9)$

3. **GRIDDED RESPONSE** How many **miles** are equivalent to 3.1 kilometers? (Use 1 kilometer $\approx$ 0.62 mile.)

4. What is the perimeter, in yards, of a rectangular garden that has a length of 10 yards and a width of 5 yards?

A. 15

B. 25

C. 30

D. 50

5. A line is graphed on the coordinate grid below.

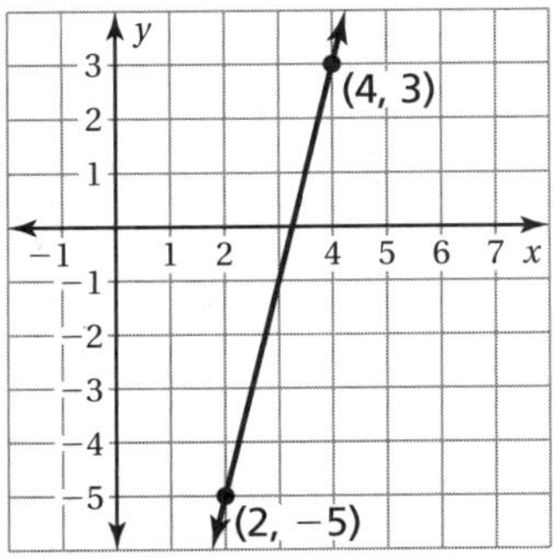

Which of the following points is found on the line?

F. (3, −2)

G. (3, 0)

H. (5, 6)

I. (5, 7)

Copyright © Big Ideas Learning, LLC
All rights reserved.

6. Which graph represents the solution of the inequality below?

$$15 \leq -3x$$

A. 

B. 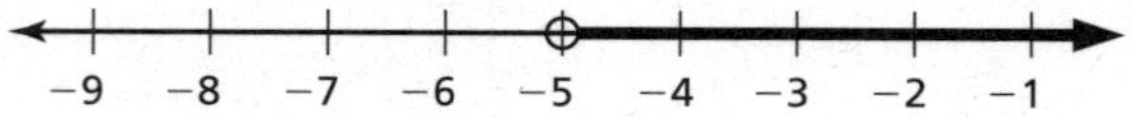

C. 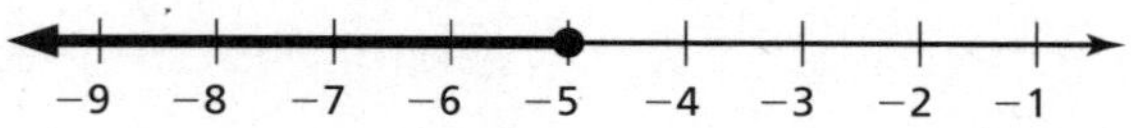

D. 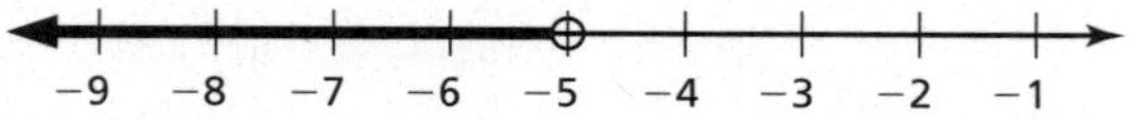

7. Which of the following describes a correct method for solving the equation below?

$$\frac{4}{9} + z = 30$$

F. Add $\frac{4}{9}$ to each side.

G. Subtract $\frac{4}{9}$ from each side.

H. Multiply each side by $\frac{4}{9}$.

I. Multiply each side by $\frac{9}{4}$.

8. What are the two numbers that could go in the box to make the equation below true?

$$|6 + \square| = 4$$

A. -4 and 4

B. -2 and 2

C. 2 and 10

D. -2 and -10

9. **EXTENDED RESPONSE** The school band decided to raise money by holding a car wash. The band spent \$32 to buy the needed supplies. They will charge \$5 for each car that they wash.

Part A Create a table showing the profit (total money earned minus expenses) that the band will make for washing 1 through 10 cars. Explain your reasoning.

Part B What will be the band's profit for washing 40 cars? Explain your reasoning.

Profit for 40 cars \$ ______________________

Part C At some point during the day, could the band ever have a profit of exactly \$107? Explain your reasoning.

Copyright © Big Ideas Learning, LLC
All rights reserved.

Chapter 14

Standards Assessment Item Analysis

1. **A.** The student incorrectly thinks that adding the same amount to two quantities keeps them in an equivalent ratio.

   B. The student incorrectly thinks that subtracting the same amount from two quantities keeps them in an equivalent ratio.

   C. Correct answer

   D. The student finds an equivalent ratio on the left side of the equation, but doubles the right side of the equation while not doubling the left.

2. **F.** The student realizes that answer choices H and I are negative, but thinks that addition must be greater than subtraction.

   G. Correct answer

   H. The student chooses the answer that has the greatest magnitude.

   I. The student thinks that -1 is greater than 0 (answer choice F) and also thinks that $9 - (-9) = -18$.

3. Correct answer: 1.922

   Common error: The student finds how many kilometers are equal to 3.1 miles, getting 5.

4. **A.** The student finds only half the perimeter of the rectangular garden.

   B. The student finds the area of a triangular garden with a base of 10 yards and a height of 5 yards.

   C. Correct answer

   D. The student finds the area of the rectangular garden.

5. **F.** The student does not apply the concept of slope correctly and chooses a point by visual means only.

   G. The student does not apply the concept of slope correctly and chooses a point by visual means only.

   H. The student does not apply the concept of slope correctly and chooses a point by visual means only.

   I. Correct answer

6. **A.** The student shades in the wrong direction, possibly forgetting to reverse the inequality symbol after dividing both sides by a negative number.

   B. The student uses the incorrect type of circle. The student also shades in the wrong direction, possibly forgetting to reverse the inequality symbol after dividing both sides by a negative number.

   C. Correct answer

   D. The student uses the incorrect type of circle.

Copyright © Big Ideas Learning, LLC
All rights reserved.

Standards Assessment Item Analysis

(continued)

7. **F.** The student incorrectly thinks that the inverse operation of addition is addition.
   G. Correct answer
   H. The student incorrectly thinks that the inverse operation of addition is multiplication.
   I. The student incorrectly thinks that the inverse operation of addition is multiplying by the reciprocal.

8. **A.** The student ignores the 6 and finds two values that each have an absolute value of 4.
   B. The student recognizes that one of the numbers must be -2 and then concludes that the other must be its opposite.
   C. The student changes the 6 to -6 and finds two numbers that satisfy the changed equation.
   D. Correct answer

9. **4 points** The student demonstrates a thorough understanding of interpreting a problem with a starting cost and a linear rate of change. In Part A, the student correctly completes a table like the one below.

Cars washed	1	2	3	4	5	6	7	8	9	10
Profit (dollars)	–27	–22	–17	–12	–7	–2	3	8	13	18

In Part B, the student correctly determines that 40 cars would give a profit of \$168. In Part C, the student correctly recognizes that 27 cars would give a profit of \$103 and that 28 cars would give a profit of \$108, and therefore the band could never have a profit of \$107. The student provides clear and complete work in Part B and explanations of the reasoning used in all parts.

3 points The student demonstrates an understanding of interpreting a problem with a starting cost and a linear rate of change, but the student's work and explanations demonstrate an essential but less than thorough understanding.

2 points The student demonstrates a partial understanding of interpreting a problem with a starting cost and a linear rate of change. The student's work and explanations demonstrate a lack of essential understanding.

1 point The student demonstrates a limited understanding of interpreting a problem with a starting cost and a linear rate of change. The student's response is incomplete and exhibits many flaws.

0 points The student provided no response, a completely incorrect or incomprehensible response, or a response that demonstrates insufficient understanding of interpreting a problem with a starting cost and a linear rate of change.

Copyright © Big Ideas Learning, LLC
All rights reserved.

Name______________________________ Date__________

Chapter 14 Alternative Assessment

1. The Film Club raised $90 with a bake sale. The members want to buy DVDs for their film library. Two members went online and found three stores that were having sales. They summarized the information in a table.

	Last Gasp	Greener Gadgets	Tech Things
1 DVD	$12	$8.95	$7.95
3 DVDs	$30	$23.85	$23.85
5 DVDs	$40	$34.75	$39.75
10 DVDs	$60	$59.50	$79.50
Shipping	$1 per DVD	$1.99 per DVD	Free
Handling	$1 per DVD	$0.99 per DVD	Free

a. Find the unit rate of cost per DVD for each offer at each store. Make a table to summarize your data.

b. The club decides to buy 9 DVDs. The unit price for 9 DVDs is the same as the unit price for 5 DVDs at all three stores. Find the price of 9 DVDs at each store. Which store offers the best buy?

c. Now calculate the total cost, including shipping and handling, for 9 DVDs at the three stores. Which store offers the lowest total cost?

2. Morgan and Terry invented a game and named it *Ratio Battle*. Each player has a set of 13 ratio cards with these ratios.

$\frac{2}{12}, \frac{4}{12}, \frac{6}{12}, \frac{8}{12}, \frac{10}{12}, \frac{11}{12}, \frac{1}{6}, \frac{2}{6}, \frac{3}{6}, \frac{4}{6}, \frac{5}{6}, \frac{1}{3}, \frac{2}{3}$

Rules: Players shuffle their sets and turn them facedown. On each round, they turn over the top cards and compare the ratios. The player with the greater ratio wins the round and keeps both cards in a separate stack. If there is a tie, both players move their card into the separate stack.

a. The table shows the first 12 rounds. Who won the most rounds? Explain.

Morgan	$\frac{4}{12}$	$\frac{1}{3}$	$\frac{3}{6}$	$\frac{2}{3}$	$\frac{8}{12}$	$\frac{6}{12}$	$\frac{11}{12}$	$\frac{2}{12}$	$\frac{1}{6}$	$\frac{2}{6}$	$\frac{4}{6}$	$\frac{5}{6}$
Terry	$\frac{2}{3}$	$\frac{2}{12}$	$\frac{5}{6}$	$\frac{10}{12}$	$\frac{4}{6}$	$\frac{2}{6}$	$\frac{6}{12}$	$\frac{1}{6}$	$\frac{1}{3}$	$\frac{3}{6}$	$\frac{4}{12}$	$\frac{11}{12}$

b. Decide who will win the last round and who will win the game. Explain your reasoning.

Copyright © Big Ideas Learning, LLC
All rights reserved.

Name ______________________________ Date __________

Chapter 14 Alternative Assessment Rubric

Score	Conceptual Understanding	Mathematical Skills	Work Habits
4	Shows complete understanding of: • finding a unit rate • evaluating expressions • comparing ratios	Calculates all unit prices accurately. Analyzes all relationships between number of DVDs and total cost correctly. Uses cross products or equivalent fractions accurately to compare all ratios.	Answers all parts of each problem. Tables are made carefully and accurately. Work is neat and well organized.
3	Shows nearly complete understanding of: • finding a unit rate • evaluating expressions • comparing ratios	Calculates most unit prices accurately. Analyzes some relationships between number of DVDs and total cost correctly. Uses cross products or equivalent fractions accurately to compare most ratios.	Answers all parts of each problem. Some mistakes are made in making or copying a table. Work is neat and easy to follow.
2	Shows some understanding of: • finding a unit rate • evaluating expressions • comparing ratios	Calculates some unit prices accurately. Does not analyze all relationships between number of DVDs and total cost correctly. Uses cross products or equivalent fractions accurately to compare some ratios.	Answers all parts of each problem. Many mistakes are made in making or copying a table. Work is sloppy and hard to follow.
1	Shows little understanding of: • finding a unit rate • evaluating expressions • comparing ratios	Does not calculate unit prices accurately. Cannot analyze relationships between number of DVDs and total cost. Does not use cross products or equivalent fractions to compare ratios.	Does not answer all parts of each problem. The tables are not made or copied carefully or accurately. Work is sloppy and hard to follow.

Copyright © Big Ideas Learning, LLC
All rights reserved.

Name______________________________ Date__________

Chapter 15 Quiz

For use after Section 15.4

Write the percent as a decimal.

1. 73% **2.** 32.3% **3.** 121.6%

Write the decimal as a percent.

4. 0.65 **5.** 1.35 **6.** 0.003

Use a number line to order the numbers from least to greatest.

7. $74\%, \frac{3}{4}, 0.76$

8. $26.2, 262\%, 2\frac{3}{5}$

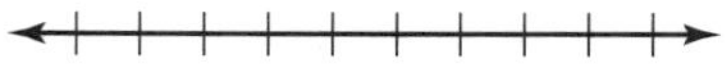

9. The table shows the scores for three students on a math test. Who earned the highest grade?

Student	Score
Jim	$\frac{17}{20}$
Sarah	84%
Ben	0.86

Write and solve a proportion to answer the question.

10. What number is 80% of 65?

11. 32 is what percent of 200?

Write and solve an equation to answer the question.

12. 125 is what percent of 200?

13. 45 is 90% of what number?

14. You earned a 90% on a science test. You answered 18 questions correctly. How many questions were on the test?

15. The table shows the percentage of total athletes at a school that play certain sports. Each athlete only plays one sport. Twenty athletes play baseball. How many athletes are there at the school?

Sport	Baseball	Football	Other
Percentage	?%	30%	45%

Answers

1. ____________
2. ____________
3. ____________
4. ____________
5. ____________
6. ____________
7. See left.
8. See left.
9. ____________
10. ____________ ____________
11. ____________ ____________
12. ____________ ____________
13. ____________ ____________
14. ____________
15. ____________

Copyright © Big Ideas Learning, LLC
All rights reserved.

Name ______________________________ Date __________

Chapter 15 Quiz

For use after Section 15.7

Identify the percent of change as an *increase* or a *decrease*. Then find the percent of change. Round to the nearest tenth of a percent, if necessary.

1. 120 pounds to 180 pounds
2. 10 feet to 8 feet
3. 400 meters to 350 meters
4. 12 gallons to 36 gallons
5. You estimate that a small restaurant will serve 430 customers the first week it is open. The actual number of customers the first week is 400. Find the percent error.

Find the original price, discount, sale price, or selling price.

6. Original price: $130
   Discount: 60%
   Sale price: ?
7. Original price: $32
   Discount: ?
   Sale price: $8
8. Original price: ?
   Discount: 20%
   Sale price: $14.40
9. Cost to store: $45
   Markup: 35%
   Selling price: ?

An account earns simple interest. Find the interest earned, principal, interest rate, or time.

10. Interest earned: ?
    Principal: $1450
    Interest rate: 9%
    Time: 5 years
11. Interest earned: $10
    Principal: $250
    Interest rate: 4%
    Time: ?
12. Interest earned: $40
    Principal: $400
    Interest rate: ?
    Time: 2 years
13. Interest earned: $45
    Principal: ?
    Interest rate: 3%
    Time: 2 years
14. Store A sells a watch for $50 and offers a 5% discount. Store B sells the same watch for $60 and offers a 20% discount. From which store should you buy?
15. A store sells a television for $1000. Customers can choose to receive a 10% discount and pay it off with a loan at a simple interest rate of 4%, or they can choose to pay the full price and pay it off in 3 years with no interest. If the customer plans to pay it off in 3 years, which option is better?
16. A store offers a loan for $900 to buy a computer. The terms of the loan are for 9% simple interest and equal monthly payments for three years. What is the monthly payment?

Answers

1. ______________
   ____
2. ______________
   ____
3. ______________
   ____
4. ______________
   ____
5. ______________
6. ______________
7. ______________
8. ______________
9. ______________
10. ______________
11. ______________
12. ______________
13. ______________
14. ______________
15. ______________
16. ______________

Copyright © Big Ideas Learning, LLC
All rights reserved.

Name________________________________ Date__________

Chapter 15 Test A

Write the percent as a decimal.

Answers

1. ______
2. ______
3. ______

4. ______

5. ______

6. ______
7. ______

8. ______

9. ______

10. ______

11. ______

12. ______

13. ______
14. ______

1. 66.7% **2.** 0.1%

3. Thursday afternoon between 4:00 and 5:00 you spend $\frac{1}{4}$ of the hour reading email, 0.4 hour doing homework, and 35% of the hour doing chores. Write the tasks in order from least amount of time to greatest.

In Exercises 4–6, use the table that shows the results of a class poll.

Question	Yes	No
Do you walk to school?	18%	82%
Is there a pet in your home?	$\frac{14}{25}$	$\frac{11}{25}$
Do you play on a sports team?	0.42	0.58

4. Which category in the table shows the least portion of the class?

5. Which is less, the number of students with a pet or the number who play on a sports team?

6. $\frac{11}{50}$ of the students live within $\frac{1}{2}$ mile of the school. Is this greater than or less than the portion of students who walk to school?

Write and solve a proportion to answer the question.

7. 35% of what number is 21? **8.** 70 is what percent of 56?

Write and solve an equation to answer the question.

9. 17 is what percent of 68? **10.** What number is 16% of 80?

Identify the percent of change as an *increase* or *decrease*. Then find the percent of change. Round to the nearest tenth of a percent, if necessary.

11. 15 books to 21 books **12.** 60 cars to 24 cars

Use the percent of change to find the new amount.

13. $84 increased by 12% **14.** 820 brushes decreased by 25%

Copyright © Big Ideas Learning, LLC
All rights reserved.

Name ____________________ Date __________

Chapter 15 Test A (continued)

Find the original price, discount, sale price, or selling price.

15. Original price: $82
Discount: 10%
Sale price: ?

16. Original price: $125
Discount: ?
Sale price: $81.25

17. Original price: ?
Discount: 36%
Sale price: $32

18. Cost to store: $32
Markup: 16%
Selling price: ?

An account earns annual simple interest. Find the interest earned, principal, interest rate, or time.

19. Interest earned: $84
Principal: $600
Interest rate: 7%
Time: ?

20. Interest earned: ?
Principal: $1250
Interest rate: 3%
Time: 4 years

21. Interest earned: $39.60
Principal: ?
Interest rate: 11%
Time: 6 months

22. Interest earned: $3250
Principal: $5000
Interest rate: ?
Time: 10 years

An account earns annual simple interest. Find the balance of the account.

23. $250 at 4% for 1 year

24. $2000 at 9% for 6 months

25. The percent of sales tax is 6%. What is the sales tax on a skateboard that costs $98?

26. The price of your favorite brand of jeans was $35 last month. This month the price is $42. What is the percent of change from last month to this month?

27. You are shopping for a cell phone. At which store should you buy the cell phone? Explain your answer.

Store	Original Price	Discount
A	$129	30%
B	$135	35%
C	$150	40%

28. You deposit $200 in an account earning 3.5% simple interest. How long will it take for the balance of the account to be $221?

Answers

15. ____________
16. ____________
17. ____________
18. ____________
19. ____________
20. ____________
21. ____________
22. ____________
23. ____________
24. ____________
25. ____________
26. ____________
27. See left.
28. ____________

Copyright © Big Ideas Learning, LLC
All rights reserved.

Name______________________________ Date__________

Chapter 15 Test B

Write the percent as a decimal.

Answers

1. 0.05%

2. 128%

3. A teacher weights the final grade as follows: $\frac{1}{5}$ for homework, 25% for a project, 0.4 for quizzes, and 15% for the final exam. Order these categories from least weighted to greatest weighted.

In Exercises 4–6, use the table.

Question	6th-graders	7th-graders	8th-graders
Participate in sports?	$\frac{3}{20}$	18%	0.22
Participate in other school activities?	33%	0.45	$\frac{11}{25}$

4. Which category in the table shows the least portion?

5. Which is greater, the percent of 7th-graders or 8th-graders participating in other school activities?

6. Your friend says that at least twice as many 7th-graders participate in other activities as play sports. Is this true? Explain.

Write and solve a proportion to answer the question.

7. What number is 45% of 60?

8. 48 is what percent of 160?

Write and solve an equation to answer the question.

9. 8 is 0.5% of what number?

10. What percent of 130 is 182?

Identify the percent of change as an *increase* or *decrease*. Then find the percent of change. Round to the nearest tenth of a percent, if necessary.

11. $\frac{2}{5}$ to $\frac{4}{5}$

12. \$18.75 to \$18.60

Find the new amount.

13. 2000 miles increased by 33%

14. 140 degrees decreased by 65%

1. ____________
2. ____________
3. ____________
4. ____________
5. ____________
6. ____________
7. ____________
8. ____________
9. ____________
10. ____________
11. ____________
12. ____________
13. ____________
14. ____________

Copyright © Big Ideas Learning, LLC
All rights reserved.

Name ________________________________ Date __________

Chapter 15 Test B (continued)

Answers

15. ________
16. ________
17. ________
18. ________
19. ________
20. ________
21. ________
22. ________
23. ________
24. ________
25. ________
26. a. ________
    b. ________
27. ________
28. See left.

Find the original price, discount, sale price, or selling price.

15. Original price: $0.75
Discount: ?
Sale price: $0.15

16. Original price: ?
Discount: 30%
Sale price: $206.50

17. Original price: $24.50
Discount: 18%
Sale price: ?

18. Cost to store: $145
Markup: 150%
Selling price: ?

An account earns annual simple interest. Find the interest earned, principal, interest rate, or time.

19. Interest earned: ?
Principal: $800
Interest rate: 6.5%
Time: 2 years

20. Interest earned: $7.50
Principal: ?
Interest rate: 12%
Time: 3 months

21. Interest earned: $235.50
Principal: $7850
Interest rate: 2%
Time: ?

22. Interest earned: $837
Principal: $3100
Interest rate: ?
Time: 6 years

An account earns annual simple interest. Find the balance of the account.

23. $1400 at 1.25% for 3 years

24. $5000 at 6% for 9 months

25. The gas tank of your car is 40% full. There are 8 gallons of gas in the tank. What is the capacity of the gas tank?

26. You purchased a stock on Monday for $24 per share.

a. On Tuesday, the stock price was $36 per share. What was the percent increase?

b. On Wednesday, the stock price had a percent decrease of 50%. What was the new stock price?

27. The cost of manufacturing a printer is $35. The manufacturer has a markup of 15% when selling to a retailer. The retailer has a markup of 35%. What is the selling price of the printer?

28. How long will it take $500 to double at a simple interest rate of 5%? Explain how you found your answer.

Copyright © Big Ideas Learning, LLC
All rights reserved.

Name____ Date__________

Chapter 15 Standards Assessment

1. Sonja bought a new van for her business. The cost of the van was $35,000, and she had to pay 6% sales tax. What was the total cost of the van, including tax?

A. $32,900 **C.** $35,600

B. $35,210 **D.** $37,100

2. What is the value of the expression $\frac{3}{4} \div \left(-\frac{1}{8}\right)$?

F. -6 **H.** $\frac{1}{6}$

G. $-\frac{1}{6}$ **I.** 6

3. **GRIDDED RESPONSE** A radio station gave free tickets to 10% of the people attending a concert. The radio station gave 640 people free tickets. What is the total number of people who attended the concert?

4. The Sharadze family is planning a car trip to Atlanta, Georgia, which is 279 miles from their home. They plan to drive at an average speed of 62 miles per hour. They also plan to stop for lunch for 45 minutes. Based on their plan, when should they leave in order to arrive in Atlanta at 3:00 P.M.?

A. 9:25 A.M. **C.** 10:30 A.M.

B. 9:45 A.M. **D.** 11:15 A.M.

5. What is the slope of the line?

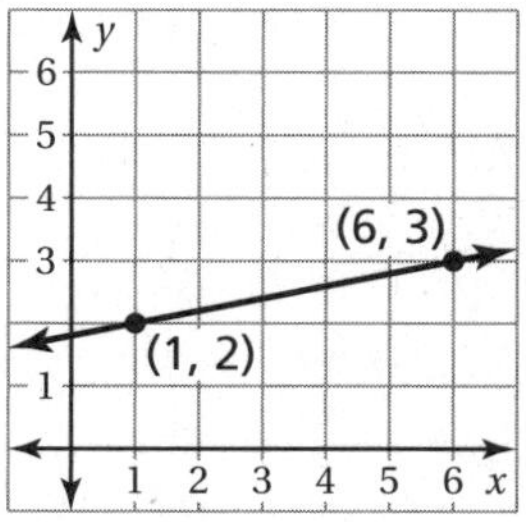

F. $\frac{1}{5}$ **H.** 2

G. $\frac{1}{2}$ **I.** 5

6. What number goes in the box below to make the statement true?

6% of ☐ is 300.

A. 18 **C.** 500

B. 180 **D.** 5000

Copyright © Big Ideas Learning, LLC
All rights reserved.

Standards Assessment (continued)

7. Two candidates ran for the office of county commissioner. They were the only two candidates who received votes in the election.

- Ms. Talmadge received 45% of the votes.
- Mr. Rodriguez received 171,919 votes.

What was the total number of votes cast in the election?

F. 140,661 **H.** 312,580

G. 249,283 **I.** 382,042

8. Which of the following describes a correct method for solving the equation below?

$$8 - y = -4$$

A. Add 8 to both sides.

B. Subtract 8 from both sides.

C. Add 8 to both sides, then add y to both sides.

D. Subtract 8 from both sides, then divide both sides by -1.

9. What is the solution of the inequality below?

$$-\frac{4}{5}x \geq \frac{1}{10}$$

F. $x \leq -\frac{2}{25}$ **H.** $x \leq -\frac{1}{8}$

G. $x \geq -\frac{2}{25}$ **I.** $x \geq -\frac{1}{8}$

10. SHORT RESPONSE Dwayne wants to buy a bowling ball that has a price of $120. As a member of a bowling league, he is entitled to a 15% discount off the price of the bowling ball. He will also have to pay 6% sales tax on the discounted price of the bowling ball.

Dwayne thinks that the final price of the bowling ball can be determined by taking 9% off the original price. Is Dwayne correct? Support your answer with a detailed explanation.

Copyright © Big Ideas Learning, LLC
All rights reserved.

Chapter 15 Standards Assessment Item Analysis

1. **A.** The student subtracts 6% of $35,000 from the cost of the van.
   B. The student makes a decimal error in calculating the tax, adding $210 to the cost of the van instead of $2,100.
   C. The student thinks that 6% is equivalent to $600 and adds this amount to the cost of the van.
   D. Correct answer

2. **F.** Correct answer
   G. The student inverts the first fraction instead of the second fraction and multiplies.
   H. The student inverts the first fraction instead of the second fraction, multiplies, and makes a sign error.
   I. The student divides correctly but makes a sign error.

3. Correct answer: 6400
   Common error: The student finds 10% of 640, getting an answer of 64.

4. **A.** The student determines the 4.5 hours of driving time but interprets this as 4 hours 50 minutes. The student then works backward from 3:00 P.M., using this driving time and the time for lunch.
   B. Correct answer
   C. The student ignores the time for lunch.
   D. The student adds the time for lunch instead of subtracting.

5. **F.** Correct answer
   G. The student just uses the point (6, 3) and finds the ratio of y to x.
   H. The student just uses the point (1, 2) and finds the ratio of y to x.
   I. The student finds the ratio of the change in x to the change in y.

6. **A.** The student finds 6% of 300.
   B. The student finds 60% of 300.
   C. The student thinks that 6% is equivalent to 0.6 and divides 300 by 0.6.
   D. Correct answer

7. **F.** The student finds the number of votes Ms. Talmadge received.
   G. The student finds 45% of 171,919 (rounded to the nearest whole number) and adds this amount to 171,919.
   H. Correct answer
   I. The student divides 171,919 by 0.45 (rounded to the nearest whole number).

Copyright © Big Ideas Learning, LLC
All rights reserved.

8. A. The student thinks that $8 - y$ is equivalent to $y - 8$.

B. The student ignores the sign to the left of y.

C. The student thinks both 8 and y are negative. The student also moves all the terms to the same side of the equal sign.

D. Correct answer

9. F. The student thinks that multiplying each side by $-\frac{4}{5}$ will solve the inequality.

G. The student thinks that multiplying each side by $-\frac{4}{5}$ will solve the inequality and forgets to reverse the inequality symbol.

H. Correct answer

I. The student forgets to reverse the inequality symbol.

10. 2 points The student demonstrates a thorough understanding of finding discounts and markups. The student thoroughly explains that the 15% discount and 6% tax are calculated on different base amounts and that it would not be equivalent to taking 9% off the original price.

1 point The student demonstrates a partial understanding of finding discounts and markups. The student is able to calculate discounts and markups but does not adequately explain why a 15% discount and 6% tax is not equivalent to taking 9% off the original price.

0 points The student demonstrates insufficient understanding of finding discounts and markups. The student is unable to calculate discounts and markups.

Copyright © Big Ideas Learning, LLC
All rights reserved.

Name______________________________ Date__________

Chapter 15 Alternative Assessment

1. Four friends are planning to buy jeans. Several stores are having sales in which they mark down items more each week as the sale goes on. The friends have checked the markdowns for the jeans they like at the three stores. One of the four friends summarized all the sale information in a table.

	Top Down Values		Base Minus Sales		Bargains All The Time	
Week	**Markdown**	**Price**	**Markdown**	**Price**	**Markdown**	**Price**
0	None	\$60	None	\$60	None	\$60
1	30% off		20% off		10% off	
2	20% more off sale price		20% more off sale price		20% more off sale price	
3	10% more off sale price		20% more off sale price		30% more off sale price	

a. Without calculating, predict which store will have the best price during the third week. Explain your reasoning.

b. Calculate the prices of jeans at each store for each week.

c. Compare and contrast the markdowns as the sales go on.

d. Which store offers the best buy each week?

e. Tell which two stores have the same week-three price for the jeans. What mathematical property explains why those week-three prices are the same?

2. Martha and Abigail live in Littleburg, where there is one bank. This bank only offers savings accounts with simple interest. Ten years ago, they both deposited \$500 in accounts paying 3% simple interest. They have not added any money to these accounts, but Abigail has done something else with her account. Look at the account balances and try to determine why Abigail has more money in her account.

Year	Martha's Account	Abigail's Account
1	\$515.00	\$515.00
2	\$530.00	\$530.45
5	\$575.00	\$579.64
10	\$650.00	\$671.96

Copyright © Big Ideas Learning, LLC
All rights reserved.

Name ______________________________ Date __________

Chapter 15 Alternative Assessment Rubric

Score	Conceptual Understanding	Mathematical Skills	Work Habits
4	Shows complete understanding of: • finding percents • calculating markdowns • simple interest	Calculates all markdowns accurately. Compares all prices correctly. Finds patterns in both pricing and interest and determines the answers to all the relevant questions.	Answers all parts of the first problem and prepares a cohesive response to the second question. Work is well organized and progresses smoothly from step to step. Work is very neat.
3	Shows nearly complete understanding of: • finding percents • calculating markdowns • simple interest	Calculates most markdowns accurately. Compares most prices correctly. Finds patterns in either pricing or interest and determines the answers to most of the relevant questions.	Answers most parts of the first problem and gives a response to the second question. Work is organized and progresses from step to step. Work is neat.
2	Shows some understanding of: • finding percents • calculating markdowns • simple interest	Calculates all markdowns accurately. Compares all prices correctly. Finds patterns in either pricing or interest and determines the answers to some of the relevant questions.	Answers some parts of the first problem and attempts to respond to the second question. Work is somewhat organized and moves from step to step. Work is very sloppy.
1	Shows little understanding of: • finding percents • calculating markdowns • simple interest	Does not calculate markdowns accurately. Does not compare prices correctly. Does not find a pattern in either pricing or interest and does not determine the answers to any of the relevant questions.	Answers some parts of the first problem and does not attempt to respond to the second question. Work is not organized and does not progress from step to step. Work is very sloppy.

Copyright © Big Ideas Learning, LLC
All rights reserved.

Name________________________________ Date__________

Test 1

End-of-Course Test

Simplify the expression. Identify the properties used.

1. $4(x - 3)$
2. $(3 \bullet x) \bullet 7$

3. Use a formula to find the area of the figure.

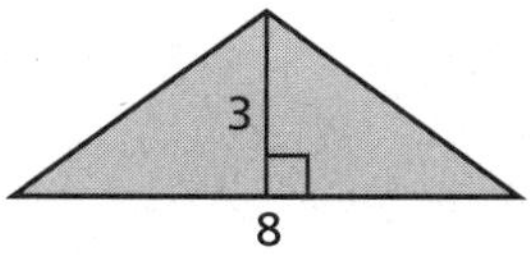

4. Tickets to a basketball game cost \$4 for adults and \$2 for children. Write an expression that gives the total cost for a adults and c children to attend the game. What is the total cost for a family of 2 adults and 3 children to attend the game?

Perform the indicated operation.

5. $\frac{7}{9} \times \frac{6}{5}$
6. $2\frac{1}{4} \div \frac{3}{8}$
7. 0.35×1.2
8. $0.25\overline{)7.38}$

9. Find the area of the parallelogram.

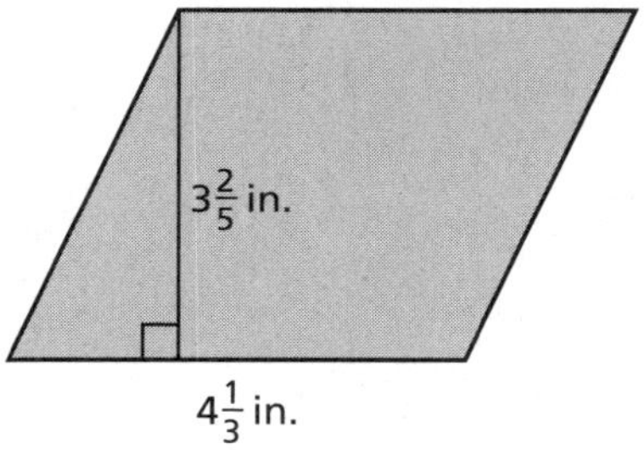

Evaluate the expression.

10. $6 + 9 \div 3$
11. $5^2 - 4 \times 2$
12. $(4 - 2)^3 - 2(3 + 1)$
13. $15 + 3(6 \div 2) - 4^2$

14. A recipe for a batch of 3 dozen chocolate chip cookies calls for 3 cups of flour, 1 cup of sugar, and 2 cups of chocolate chips. How much of each ingredient should be used to make 2 dozen cookies?

15. Find the area of the polygon with vertices of $A(0, 1)$, $B(0, 5)$, $C(4, 5)$, and $D(6, 1)$.

16. The ages of people on a jury are 41, 45, 39, 56, 48, 45, 42, 34, 47, 62, 35, and 58. Make a stem-and-leaf plot of the data.

Answers

1. ______________
____
2. ______________
____
____
3. ______________
4. ______________
____
5. ______________
6. ______________
7. ______________
8. ______________
9. ______________
10. ______________
11. ______________
12. ______________
13. ______________
14. ______________
____
____
15. ______________
16. See left.

Copyright © Big Ideas Learning, LLC
All rights reserved.

Name ______________________________ Date __________

Test 1 End-of-Course Test (continued)

Order the integers from least to greatest.

17. 7, 3, −2, −4, 5

18. −5, −1, 3, 0, −3

19. A twelve-pack of juice costs $4.20. An eighteen-pack costs $5.40. Which is the better buy?

Write the fraction or mixed number as a percent.

20. $\frac{3}{8}$

21. $\frac{6}{5}$

22. $3\frac{1}{4}$

23. Chris, Mary Beth, and Allison are discussing the number of oranges grown in Florida. Chris says that approximately 14.6% of the world's oranges are grown in Florida, Mary Beth says that 292 out of every 2000 oranges are grown in Florida, and Allison says that 0.146 of the world's oranges are grown in Florida. Are they in agreement? Explain your reasoning.

24. How many vertices does a triangular prism have?

25. A pizza shop offers 30% off the price of a large pizza every Tuesday night. If the regular price is $25, what is the discounted price?

26. Write the ratio of basketballs to footballs. Explain what the ratio means.

27. You run 6 miles in 1 hour. At this rate, how long will it take you to run a marathon (approximately 26 miles)?

28. Determine the mean, median, mode(s), IQR, and range for the data.

3, 8, 6, 6, 4, 6, 9, 9, 12

29. Katie makes 70% of her shots from the free-throw line. Can you determine how many consecutive free-throws she must make in order to increase her percentage to 75%? Explain.

Answers

17. ____________

18. ____________

19. ____________

20. ____________

21. ____________

22. ____________

23. ____________

24. ____________

25. ____________

26. ____________

27. ____________

28. ____________

29. ____________

Copyright © Big Ideas Learning, LLC
All rights reserved.

Name______________________________ Date__________

Test 1

End-of-Course Test (continued)

Plot the ordered pair in the coordinate plane.

30. $(3, -4)$ **31.** $(-4, 2)$

32. $(-2, 0)$ **33.** $(-3, -3)$

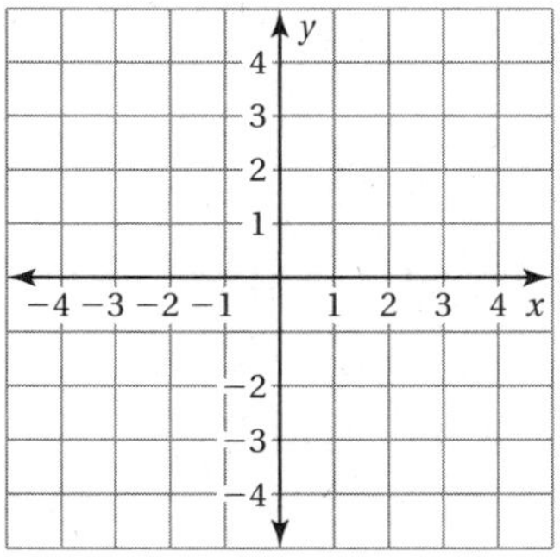

Solve the equation.

34. $s + 3 = 13$ **35.** $4c = 24$

36. A farmer builds a fence to enclose a rectangular pasture. He uses 160 feet of fence. Find the total area of the pasture if it is 50 feet long.

37. Write and solve an equation to find the width of the box if its volume is 96 cubic centimeters. Then find its surface area.

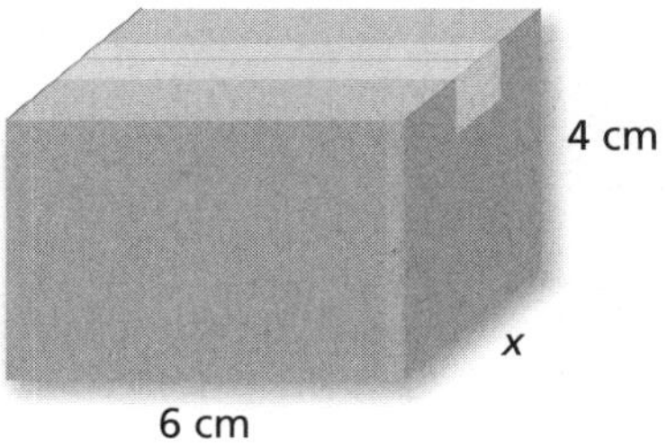

38. The prices of backpacks at a store are \$22, \$16, \$39, \$35, \$19, \$34, \$20, and \$26. Find the mean absolute deviation of the prices.

Write the word sentence as an inequality.

39. A number t is less than 7. **40.** A number m is at least -3.

Determine whether the question is a statistical question. Explain.

41. How tall are sixth grade students in the United States?

42. How many inches are in one foot?

Use the box-and-whisker plot to answer the question.

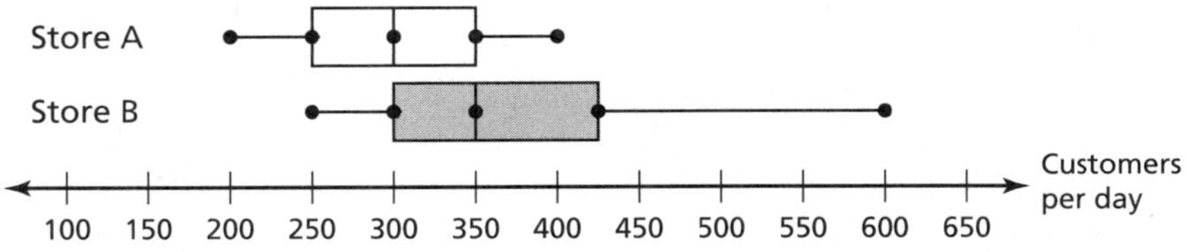

43. How often does Store A have 300 or less customers per day?

44. Identify the shape of each distribution.

45. Which store has more customers?

Answers

30. See left.
31. See left.
32. See left.
33. See left.
34. ______
35. ______
36. ______
37. ______
38. ______
39. ______
40. ______
41. ______
42. ______
43. ______
44. ______
45. ______

Copyright © Big Ideas Learning, LLC
All rights reserved.

Name ____________________ Date __________

Test 1 End-of-Course Test (continued)

Find the GCF of the numbers.

46. 30, 105

47. 84, 90

48. You have piano lessons every seventh day and cooking lessons every fourth day. Today you have both lessons. In how many days will you have both lessons on the same day again?

49. The surface area of a square pyramid is 95 square inches. The side length of the base is 5 inches. What is the value of x?

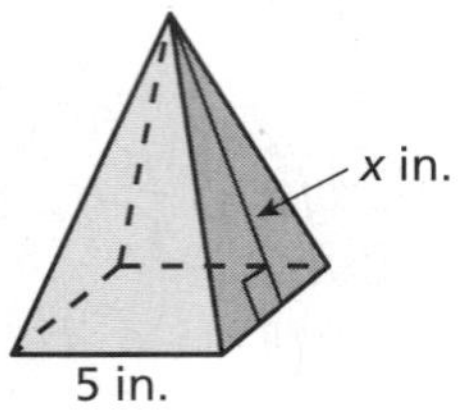

Answers

46. __________
47. __________
48. __________
49. __________
50. See left.
51. See left.
52. __________
53. __________
54. __________

Find the missing values in the ratio table.

50.

Teachers	2	3	
Students	24		120

51.

Cups	52	36	
Quarts	13		5

In Exercises 52–54, use the histogram that shows the number of school activities that students are involved in during the year.

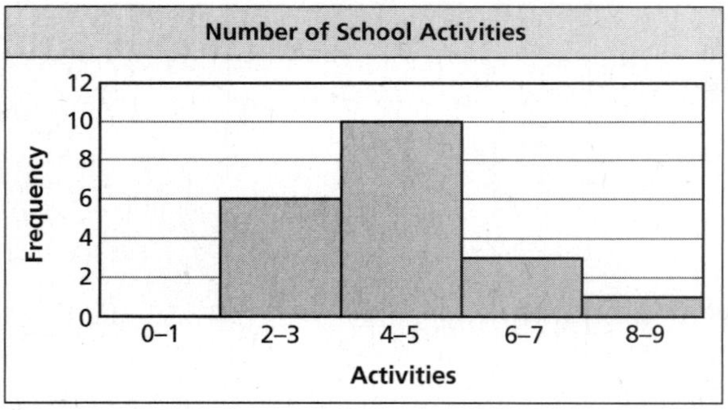

52. Which interval contains the fewest data values?

53. How many students are there?

54. Determine the percent of students that are involved in at least 4 or 5 activities.

Copyright © Big Ideas Learning, LLC
All rights reserved.

Name ______________________ Date __________

Test 1 End-of-Course Test (continued)

Order the numbers from least to greatest.

Answers

55. $|-3|, 4, -4, -|2|, -1$

56. $\frac{21}{2}, -7.5, -\frac{36}{5}, 9.5$

Simplify the expression.

57. $4 - (-3)$

58. $-2 + 15$

59. $-3(4)$

60. $27 \div (-3)$

61. $-\frac{1}{6} + \frac{7}{12}$

62. $0.24 - 1.6$

63. $2\frac{3}{5} \bullet \left(-\frac{4}{3}\right)$

64. $-24 \div 3.2$

65. On an exam you get two points for each question answered correctly, zero points for each question left blank, and lose one point for each question answered incorrectly. What is your total score on the exam if you answer 22 questions correctly, leave 7 questions blank, and answer 5 questions incorrectly?

Solve.

66. $x + 2\frac{4}{5} = 3\frac{1}{6}$

67. $-0.4a + 1.2 = 3.6$

68. A pencil costs \$0.30 and a pen costs \$0.50. You buy 10 pencils and the total cost is \$7.50. How many pens did you buy?

69. A farmer builds a fence to enclose a rectangular pasture. He uses 160 feet of fence. Find the total area of the pasture if it is 50 feet long.

70. The table shows the time in minutes m to download s songs. How long does it take to download one song?

Minutes	1	3	5
Songs	2	6	10

55. ____________
56. ____________
57. ____________
58. ____________
59. ____________
60. ____________
61. ____________
62. ____________
63. ____________
64. ____________
65. ____________
66. ____________
67. ____________
68. ____________
69. ____________
70. ____________

Copyright © Big Ideas Learning, LLC
All rights reserved.

Name ________________________________ Date __________

Test 1 End-of-Course Test (continued)

Tell whether the ratios form a proportion.

71. $\frac{3}{8}, \frac{13}{40}$

72. $\frac{7}{9}, \frac{28}{36}$

73. Solve the proportion $\frac{7}{5} = \frac{21}{x}$.

Find the slope of the line.

74.

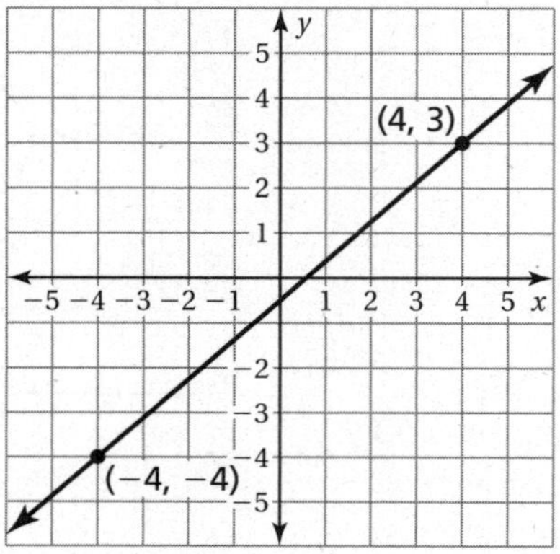

75.

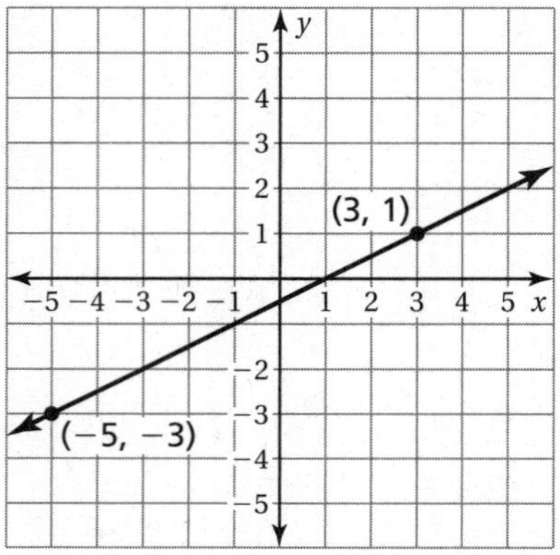

76. If 30% of a number is 15, what is the number?

77. A store sign reads "Take 75% off the original price when you take an additional 15% off the sale price, which is 60% off the original price." Is the store's sign accurate? Explain.

78. You put $1200 in an account that earns 3% simple interest. Find the total amount in the account after four years.

Answers

71. ____________

72. ____________

73. ____________

74. ____________

75. ____________

76. ____________

77. See left.

78. ____________

Copyright © Big Ideas Learning, LLC
All rights reserved.

Name ______________________ Date __________

Test 2

End-of-Course Test

Simplify the expression. Identify the properties used.

1. $4(x + 3)$

2. $(3.5 \bullet x) \bullet 4$

3. Use a formula to find the area of the figure.

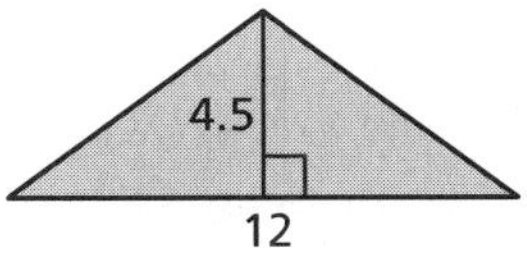

4. Tickets to a basketball game cost $3.50 for adults and $2 for children. Write an expression that gives the total cost for *a* adults and *c* children to attend the game. What is the total cost for a family of 2 adults and 3 children to attend the game?

Perform the indicated operation.

5. $\frac{3}{7} \times \frac{4}{6}$

6. $4\frac{3}{5} \div \frac{1}{8}$

7. 0.45×3.2

8. $0.35\overline{)1.61}$

9. Find the area of the parallelogram.

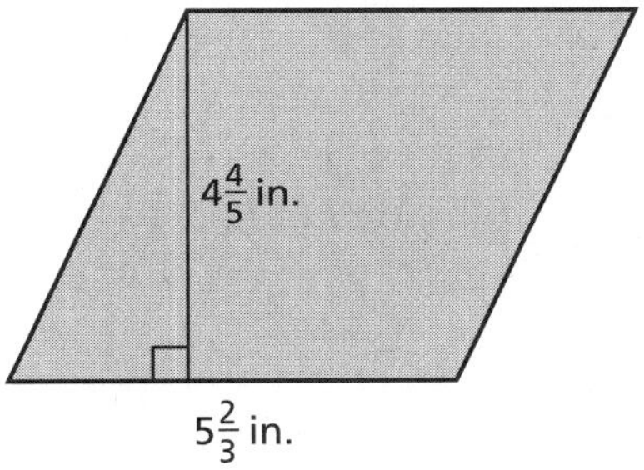

Evaluate the expression.

10. $4 + 10 \div 2$

11. $7^2 - 3 \times 4$

12. $(8 - 5)^3 - 3(1 + 2)$

13. $7 + 3(12 \div 4) - 3^2$

14. A recipe for a batch of 3 dozen chocolate chip cookies calls for $3\frac{1}{2}$ cups of flour, 1 cup of sugar, and $2\frac{3}{4}$ cups of chocolate chips. How much of each ingredient should be used to make 2 dozen cookies?

15. Find the area of the polygon with vertices of $A(2, 2)$, $B(2, 7)$, $C(8, 7)$, and $D(4, 2)$.

16. The ages of people on a jury are 56, 52, 42, 50, 58, 61, 38, 55, 69, 66, 56, and 46. Make a stem-and-leaf plot of the data.

Answers

1. ______________
____
2. ______________
____
____
3. ______________
4. ______________
____
5. ______________
6. ______________
7. ______________
8. ______________
9. ______________
10. ______________
11. ______________
12. ______________
13. ______________
14. ______________
____
____
15. ______________
16. See left.

Copyright © Big Ideas Learning, LLC
All rights reserved.

Name ______________________ Date __________

Test 2

End-of-Course Test (continued)

Order the integers from least to greatest.

17. 9, –3, 6, –2, –5

18. –8, –2, 4, 2, –1

19. A twelve-pack of juice costs $3.90. An eighteen-pack costs $5.49. Which is the better buy?

Write the fraction or mixed number as a percent.

20. $\frac{5}{8}$

21. $\frac{21}{5}$

22. $2\frac{3}{25}$

23. Chris, Mary Beth, and Allison are discussing the number of votes received by a candidate running for office. Chris says that approximately 46.2% of the votes went to the candidate, Mary Beth says that 231 out of every 500 votes went to the candidate, and Allison says that 0.462 of the votes went to the candidate. Are they in agreement? Explain your reasoning.

24. How many vertices does a rectangular pyramid have?

25. A pizza shop offers 30% off the price of a large pizza every Tuesday night. If the regular price is $25.50, what is the discounted price?

26. Write the ratio of baseballs to soccer balls. Explain what the ratio means.

27. You run 5 miles in 1 hour. At this rate, how long will it take you to run a marathon (approximately 26 miles)?

28. Determine the mean, median, mode(s), IQR, and range for the data.

12, 12, 10, 8, 9, 9, 9, 11, 11, 8

29. Katie makes 65% of her shots from the free-throw line. Can you determine how many consecutive free-throws she must make in order to increase her percentage to 68%? Explain.

Answers

17. ______
18. ______
19. ______
20. ______
21. ______
22. ______
23. ______
24. ______
25. ______
26. ______
27. ______
28. ______
29. ______

Copyright © Big Ideas Learning, LLC
All rights reserved.

Name______________________________ Date__________

Test 2

End-of-Course Test (continued)

Plot the ordered pair in the coordinate plane.

30. $(1, -3)$

31. $(-2, 4)$

32. $(0, -2)$

33. $(-4, -3)$

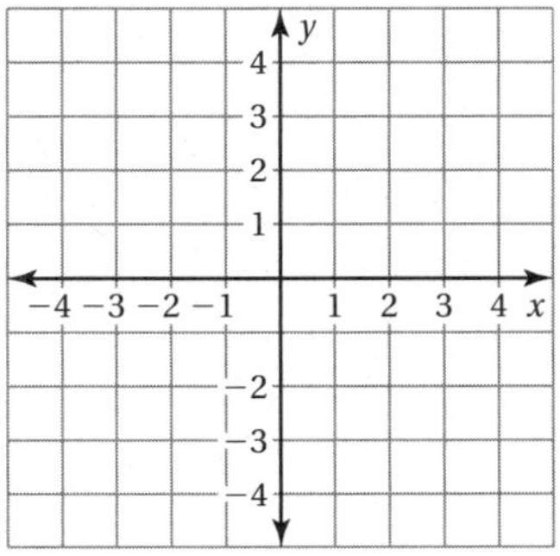

Solve the equation.

34. $\frac{3}{4}s = 12$

35. $2.5c = 20$

36. A farmer builds a fence to enclose a rectangular pasture. He uses 155 feet of fence. Find the total area of the pasture if it is 45.5 feet long.

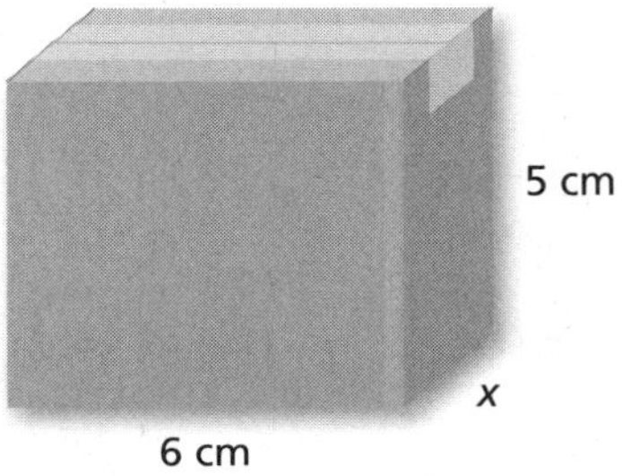

37. Write and solve an equation to find the width of the box if its volume is 80 cubic centimeters. Then find its surface area.

38. The prices of backpacks at a store are \$26, \$22, \$31, \$18, \$24, \$22, \$19, and \$30. Find the mean absolute deviation of the prices.

Write the word sentence as an inequality.

39. A number w is less than 5.5.

40. A number m is at least 7.

Determine whether the question is a statistical question. Explain.

41. How much does a 10-fluid ounce bottle of perfume cost?

42. Who was the President of the United States in 2010?

Use the box-and-whisker plot to answer the question.

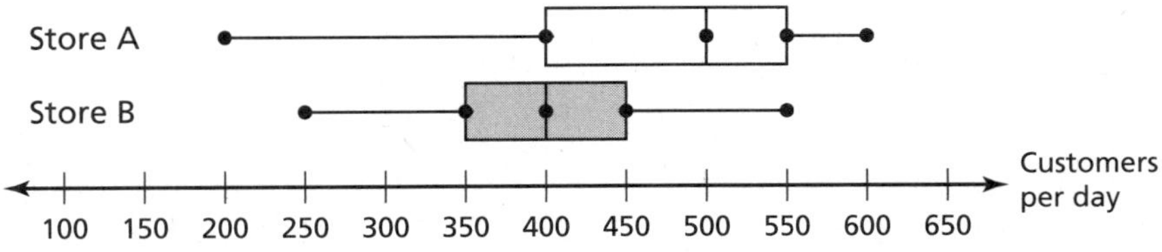

43. How often does Store A have 550 or more customers per day?

44. Identify the shape of each distribution.

45. Which store has more customers?

Answers

30. See left.
31. See left.
32. See left.
33. See left.
34. ____________
35. ____________
36. ____________
37. ____________
38. ____________
39. ____________
40. ____________
41. ____________
42. ____________
43. ____________
44. ____________
45. ____________

Copyright © Big Ideas Learning, LLC
All rights reserved.

Name ______________________ Date ________

Test 2

End-of-Course Test (continued)

Find the GCF of the numbers.

46. 20, 150

47. 42, 105

48. You have piano lessons every fifth day and cooking lessons every sixth day. Today you have both lessons. In how many days will you have both lessons on the same day again?

49. The surface area of a square pyramid is 189 square inches. The side length of the base is 7 inches. What is the value of x?

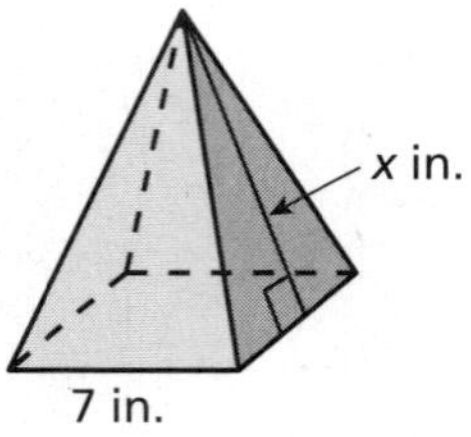

Answers

46. ________

47. ________

48. ________

49. ________

50. See left.

51. See left.

52. ________

53. ________

54. ________

Find the missing values in the ratio table.

50.

Boys	3	9	
Girls	7		42

51.

Feet	72	57	
Yards	24		13

In Exercises 52–54, use the histogram that shows the number of school activities that students are involved in during the year.

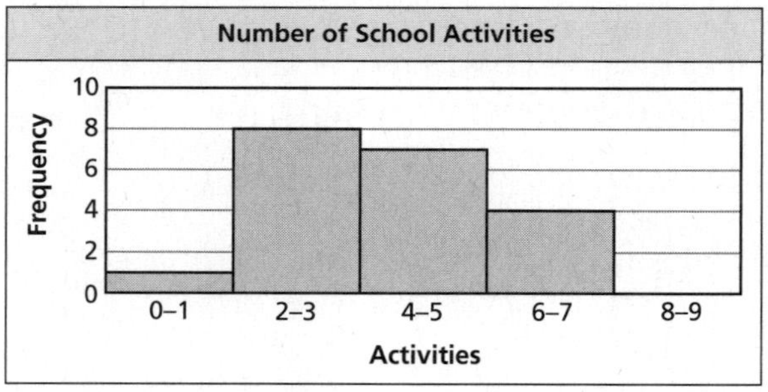

52. Which interval contains the fewest data values?

53. How many students are there?

54. Determine the percent of students that are involved in at least 4 or 5 activities.

Copyright © Big Ideas Learning, LLC
All rights reserved.

Name________________________________ Date__________

Test 2

End-of-Course Test (continued)

Order the numbers from least to greatest.

55. $-|-1|, 0, |-2|, -5, -4$

56. $\frac{17}{2}, -6.7, -\frac{28}{5}, 8.2$

Simplify the expression.

57. $2 - (-5)$

58. $-6 + 11$

59. $-2.5(6)$

60. $45 \div (-3)$

61. $-\frac{2}{5} + \frac{7}{10}$

62. $0.38 - 1.4$

63. $2\frac{1}{8} \bullet \left(-\frac{6}{5}\right)$

64. $-12.88 \div 2.8$

65. On an exam you get two points for each question answered correctly, zero points for each question left blank, and lose one-half point for each question answered incorrectly. What is your total score on the exam if you answer 13 questions correctly, leave 7 questions blank, and answer 5 questions incorrectly?

Solve.

66. $x + 3\frac{2}{5} = 2\frac{1}{6}$

67. $-0.25a + 3.8 = 4.6$

68. A pencil costs $0.27 and a pen costs $0.32. You buy six pencils and the total cost is $4.18. How many pens did you buy?

69. A farmer builds a fence to enclose a rectangular pasture. He uses 155 feet of fence. Find the total area of the pasture if it is 45.5 feet long.

70. The table shows the time in minutes m to download s songs. How long does it take to download one song?

Minutes	0.5	1.5	2.5
Songs	2	6	10

Answers

55. ____________
56. ____________
57. ____________
58. ____________
59. ____________
60. ____________
61. ____________
62. ____________
63. ____________
64. ____________
65. ____________
66. ____________
67. ____________
68. ____________
69. ____________
70. ____________

Copyright © Big Ideas Learning, LLC
All rights reserved.

Name ______________________________ Date __________

Test 2

End-of-Course Test (continued)

Tell whether the ratios form a proportion.

71. $\frac{4}{7}, \frac{12}{28}$

72. $\frac{6}{11}, \frac{30}{55}$

73. Solve the proportion $\frac{3}{8} = \frac{11}{x}$.

Find the slope of the line.

74.

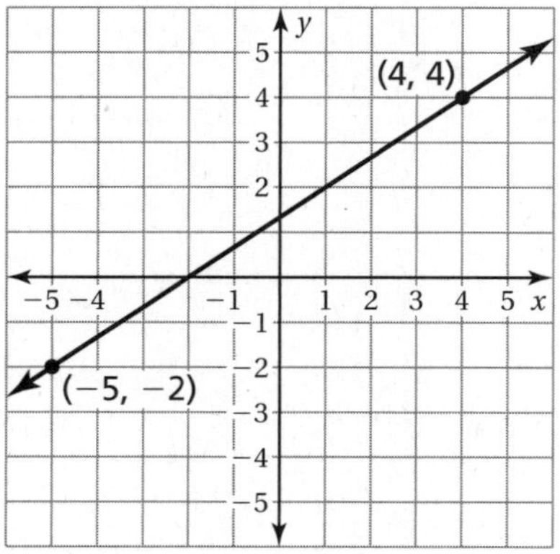

75.

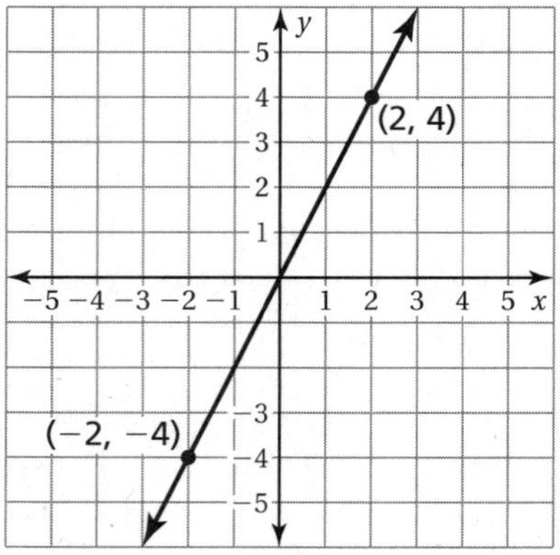

76. If 27% of a number is 11.34, what is the number?

77. A store sign reads, "Take 90% off the original price when you take an additional 15% off the sale price, which is 75% off the original price." Is the store's sign accurate? Explain.

78. You put $1205 in an account that earns 2.5% simple interest. Find the total amount in the account after four years.

Answers

71. __________

72. __________

73. __________

74. __________

75. __________

76. __________

77. See left.

78. __________

Copyright © Big Ideas Learning, LLC
All rights reserved.

Name ______________________________ Date __________

Resource Gridded Response Answer Sheet

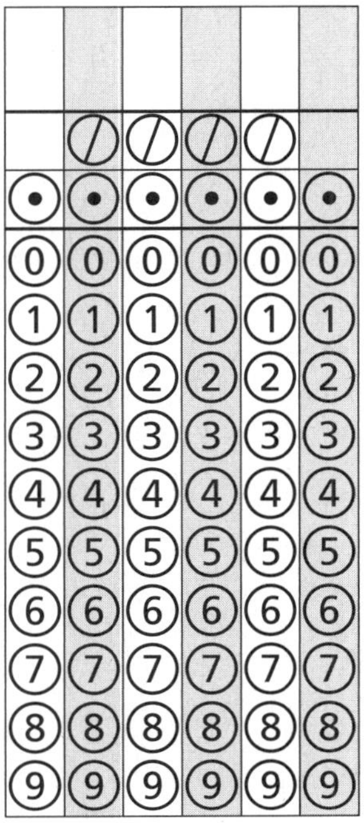

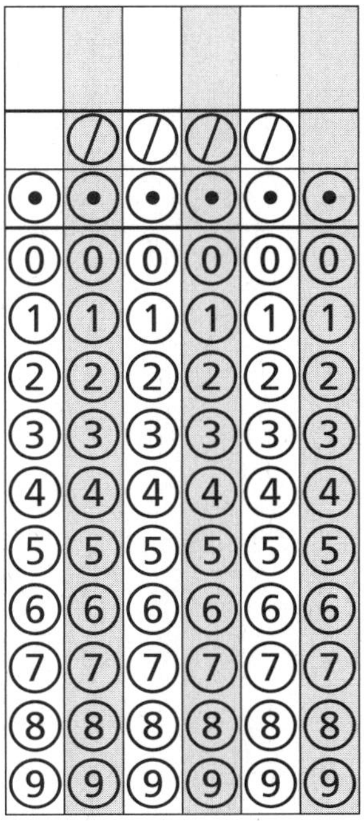

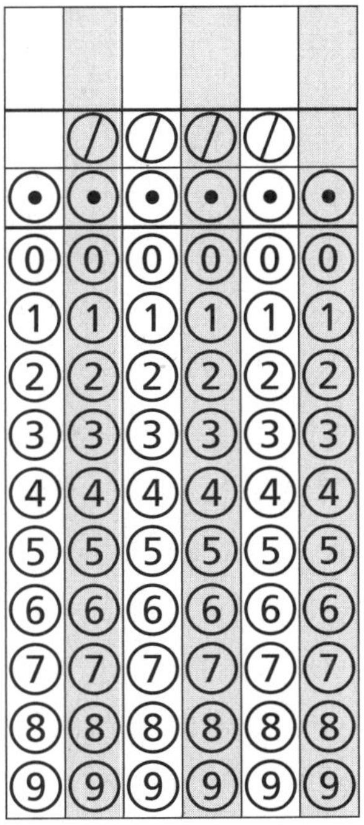

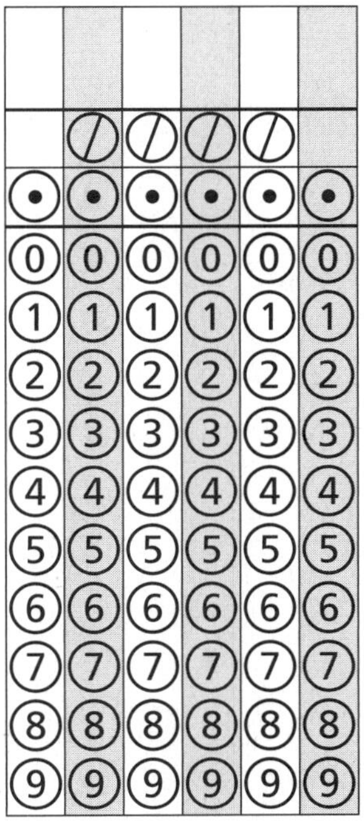

Copyright © Big Ideas Learning, LLC
All rights reserved.

Name ______________________________ Date __________

Resource Gridded Response Answer Sheet

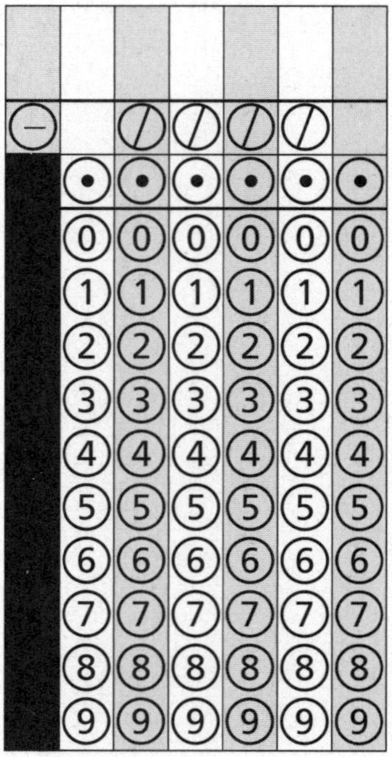

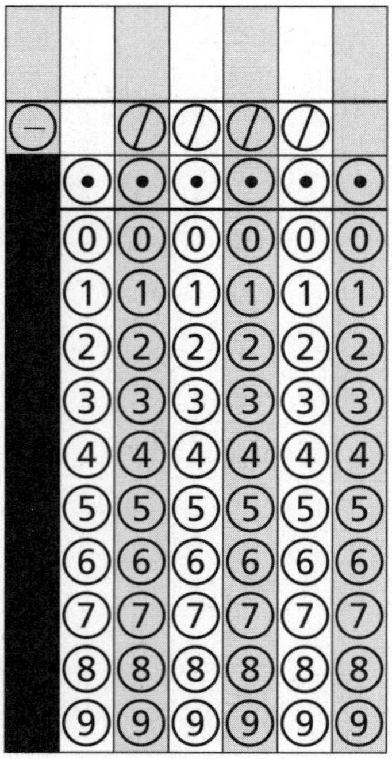

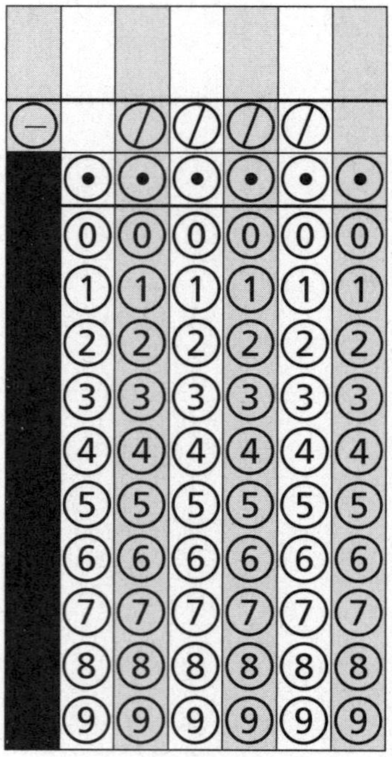

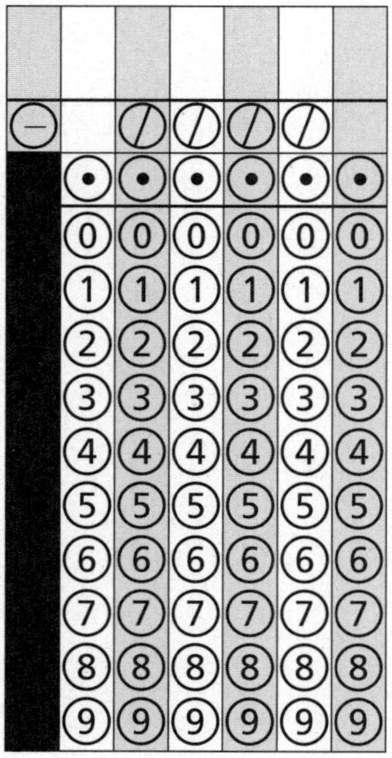

Copyright © Big Ideas Learning, LLC
All rights reserved.

Credits

7 Buki Toys LTD; **8** Adapted from Singleton, Glen. *1001 Cool Jokes*. Hinkler Books Pty Ltd., 2003; **14** Singleton, Glen. *1001 Cool Jokes*. Hinkler Books Pty Ltd., 2003; **20** Adapted from Singleton, Glen. *1001 Cool Jokes*. Hinkler Books Pty Ltd., 2003; **26** *The Youth Online Club – 101 Jokes for Kids*. [Cited 12 January 2009] Available from *www.youthonline.ca/101thingstodo/jokes1-10.shtml*; **31** Buki Toys LTD; **32** Adapted from *YAHOO Kids*. [Cited 12 January 2009] Available from *www.kids.yahoo.com/jokes*; **42** *The Youth Online Club – 101 Jokes for Kids*. [Cited 12 January 2009] Available from *www.youthonline.ca/101thingstodo/jokes61-70.shtml*; **48, 54** Keller, Charles. *Best Riddle Book Ever*. Sterling Publishing Company, 1997; **60** Singleton, Glen. *1001 Cool Jokes*. Hinkler Books Pty Ltd., 2003; **70** Weigle, Oscar. *Great Big Joke and Riddle Book*. Grosset and Dunlap, 1981; **76** Keller, Charles. *Best Riddle Book Ever*. Sterling Publishing Company, 1997; **84** Singleton, Glen. *1001 Cool Jokes*. Hinkler Books Pty Ltd., 2003; **89** Buki Toys LTD; **90** *Jokes and Funny Stories*. [Cited 12 January 2009] Available from *www.thejokes.co.uk/*; **96** Singleton, Glen. *1001 Cool Jokes*. Hinkler Books Pty Ltd., 2003; **106** *The Youth Online Club – 101 Jokes for Kids*. [Cited 12 January 2009] Available from *www.youthonline.ca/101thingstodo/jokes61-70.shtml*; **112** Singleton, Glen. *1001 Cool Jokes*. Hinkler Books Pty Ltd., 2003; **120** *National Geographic Kids*, April 2008, 15; **126** *Adapted from Kids Jokes*. [Cited 12 January 2009] Available from *www.kidsgoals.com/kids-jokes.html*; **132** Adapted from *Buzzle.com*. [Cited 12 January 2009] Available from *www.buzzle.com/articles/funny-riddles-forkids-riddles-answers.html*; **138** Adapted from Singleton, Glen. *1001 Cool Jokes*. Hinkler Books Pty Ltd., 2003; **148** Walton, Rick. *Really, Really Bad School Jokes*. Candlewick, August 1995; **154** Weitzman, Ilana, Eva Blank and Roxanne Green. *Jokelopedia: The Biggest, Best, Silliest, Dumbest, Joke Book Ever!*. Workman Publishing Company, October 2000; **160** Rosenbloom, Joseph. *Biggest Riddle Book in the World*. Sterling Publishing Company, 1976; **166** Adapted from *Man in the Moon*. [Cited 12 January 2009] Available from *www.maninthemoon.co.uk/jokes/html*; **172** Adapted from *Ducksters*. [Cited 12 January 2009] Available from *www.ducksters.com/jokesforkids/silly.php*; **178, 184** Singleton, Glen. *1001 Cool Jokes*. Hinkler Books Pty Ltd., 2003

Copyright © Big Ideas Learning, LLC
All rights reserved.

Answers for Resources by Chapter

Chapter 11

11.1 Start Thinking!

For use before Activity 11.1

Sample answer: In football, there is an order to the game. For example, one team must kick-off before the other team can begin their drive, they must score a touchdown before kicking the extra point. In math problems, there is the order of operations. Students must first multiply or divide before doing addition and subtraction.

11.1 Warm Up

For use before Activity 11.1

1. $<$ **2.** $>$ **3.** $>$

4. $<$ **5.** $>$ **6.** $<$

11.1 Start Thinking!

For use before Lesson 11.1

Sample answer: Going to school is like moving the positive direction on a number line. Returning home after school is like moving the negative direction on a number line. No, you never travel a negative direction.

11.1 Warm Up

For use before Lesson 11.1

1. 15 **2.** 23 **3.** 7 **4.** 35

5. 43 **6.** 0 **7.** 39 **8.** 212

11.1 Practice A

1. 7 **2.** 12 **3.** 13 **4.** 0

5. $>$ **6.** $=$ **7.** $>$ **8.** $-5, 3$

9. $-1, |-1|, |4|, |5|, 8$ **10.** $0, |2|, |3|, |5|, 6$

11. 19 **12.** -8 **13.** -13

14. a. 2 **b.** 2 ft per sec **c.** positive **d.** 2 ft per sec

15. a. LATE **b.** TEAL **16.** *Sample answer*: -5

11.1 Practice B

1. $=$ **2.** $<$ **3.** $<$ **4.** $15, -6$

5. $-|-34|, |0|, 14, |-25|, 28$

6. $-16, 10, |-16|, |25|, |-43|$

7. 249 **8.** -183 **9.** -153

10. a. Phosphorus **b.** Oxygen

11. a. up **b.** 13 ft/sec **c.** down **d.** 17 ft/sec

12. 0

13. true; Both numbers have an absolute value of 3.

14. false; *Sample answer:* Let $x = -4$. Then $|x| = 4$ and 4 is not less than -4.

11.1 Enrichment and Extension

1. always **2.** never **3.** sometimes

4. sometimes **5.** never **6.** always

7. sometimes **8.** sometimes

9. all negative integers **10.** all positive integers

11. none **12.** all integers **13.** none

14. all integers **15.** all integers **16.** none

17. all integers **18.** none

19. all negative integers and zero

20. all negative integers and zero

21. all positive integers and zero

22. all positive integers and zero

23. all positive integers and zero

24. all negative integers and zero

25. none **26.** all negative integers

27. all positive integers **28.** all positive integers

29. none **30.** all integers

31. all positive integers **32.** all integers

33. none **34.** all integers

35. all integers **36.** all integers

Check students' picture; It should be a dog.

11.1 Puzzle Time

A TENNIS BALL

11.2 Start Thinking!

For use before Activity 11.2

Sample answer: In golf, the goal is to get below par. This would represent zero. Scores above a par on a hole would be positive integers and scores below par on a hole would be negative integers.

11.2 Warm Up

For use before Activity 11.2

1. 22 **2.** 42 **3.** 77

4. 150 **5.** 122 **6.** 221

Copyright © Big Ideas Learning, LLC
All rights reserved.

Answers for Resources by Chapter

11.2 Start Thinking!

For use before Lesson 11.2

less than; *Sample answer:* The starting temperature increases $(\text{add } +10)$ 10 degrees, then decreases $(\text{add } -12)$ 12 degrees.

11.2 Warm Up

For use before Lesson 11.2

1. 13 **2.** −6 **3.** 5

4. 4 **5.** −4 **6.** 0

11.2 Practice A

1. 10 **2.** −8 **3.** −12 **4.** 0

5. 0 **6.** 3 **7.** 3 **8.** −6

9. −17 **10.** −17 **11.** −7 **12.** −3

13. \$29

14. Use the Commutative Property to switch the positions of the terms −5 and −8. Then use the Associative Property to group the terms 8 and −8. Because they are opposites, their sum will be zero; −5

15. Use the Commutative Property to switch the positions of the terms 4 and 9. Then use the Associative Property to group the terms −4 and 4. Because they are opposites, their sum will be zero; 9

16. Use the Commutative Property to switch the positions of the terms 12 and −7. Then use the Associative Property to group the terms −5 and −7. The sum of −5 and −7 is −12, which is the opposite of 12; 0

17. 10 **18.** −9 **19.** 11

20. 4 **21.** −7 **22.** −3

23. −18 **24.** 15 **25.** −27

26. $n = 13$ **27.** $c = -4$ **28.** $k = -8$

29. $2 + (-1) + (-2)$

30.

3	−1	−2
−4	4	0
1	−3	2

11.2 Practice B

1. 39 **2.** −48 **3.** 0

4. −41 **5.** 5

6. Use the Commutative Property to switch the positions of the terms −25 and −18. Then use the Associative Property to group the terms 18 and −18. Because they are opposites, their sum will be zero; −25

7. Use the Commutative Property to switch the positions of the terms 45 and −8. Then use the Associative Property to group the terms −22 and −8; 15

8. Use the Commutative Property to switch the positions of the terms −12 and 4. Then use the Associative Property to group the terms 28 and 4; 20

9. −28 **10.** 3 **11.** −27

12. 56 **13.** 18 **14.** −79

15. 37 **16.** 14 **17.** −59

18. $n = 25$ **19.** $c = 71$ **20.** $k = -80$

21. *Sample answer:* −30, 8, 2 and 8, 7, −5

22. −10°F

23. *Sample answer:*

9	−6	−3
−1	0	1
−8	6	2

11.2 Enrichment and Extension

1.

−8	−3	−4
−1	−5	−9
−6	−7	−2

2.

−2	3	−4
−3	−1	1
2	−5	0

3.

−7	7	6	−4
4	−2	−1	1
0	2	3	−3
5	−5	−6	8

Copyright © Big Ideas Learning, LLC
All rights reserved.

4.

−4	1	−10	3
−9	2	−3	0
5	−8	−1	−6
−2	−5	4	−7

5. Check students' work.

11.2 Puzzle Time

IN CASE HE GOT A HOLE IN ONE

11.3 Start Thinking!

For use before Activity 11.3

Sample answer: You could find the distance from where you are starting to where you are going. You could find the change in elevation.

11.3 Warm Up

For use before Activity 11.3

1. 34 2. 64 3. 77
4. 7 5. 45 6. 2

11.3 Start Thinking!

For use before Lesson 11.3

Sample answer: Rewrite the problem as an addition problem (add the opposite), then use additional rules. Do this for all 3 questions.

11.3 Warm Up

For use before Lesson 11.3

1. −3 2. 15 3. 1
4. −10 5. −7 6. −17

11.3 Practice A

1. −5 2. 9 3. −10 4. −3
5. 19 6. 8 7. −22 8. 8
9. 2 10. 22 11. 30 12. −45
13. −13 14. $7 + (-3)$ 15. $5 - 3$
16. 2 17. 13 18. −4 19. 14
20. −1 21. 0 22. −18 23. 19
24. 26 25. 10 26. −11 27. −4
28. *Sample answer:* $x = -1$, $y = -3$; $x = -4$, $y = -6$
29. a. 367 ft b. 33 ft c. New Orleans

11.3 Practice B

1. −5 2. 29 3. −49 4. −15
5. 191 6. −146 7. −12 8. 18
9. $8 - (-28)$ 10. 9
11. 17 12. 13 13. 134 14. −48
15. −44 16. −25 17. 103 18. 206
19. 71 20. 17 21. 23 22. 7
23. −38 24. −53
25. a. 94, 103, 114, 107, 84, 76, 64, 65, 75, 86, 105, 98
b. high of 99, low of −46 c. 145
26. For $|b| > |a|$ or a and b have different signs.

11.3 Enrichment and Extension

1. highest: 14; first dart: 7; second dart: −7
2. lowest: −14; first dart: −7; second dart: 7
3.

First Dart	7	6	5	4	3	2
Second Dart	3	2	1	0	−1	−2

First Dart	1	0	−1	−2	−3
Second Dart	−3	−4	−5	−6	−7

4.

First Dart	1	0	−1	−2	−3
Second Dart	7	6	5	4	3

First Dart	−4	−5	−6	−7
Second Dart	2	1	0	−1

5. There are 15 ways to get a score of 0. There is 1 way to get a score of 14. There is 1 way to get a score of −14.
6. You should try to land on a negative integer first and a positive integer second. Subtracting a positive integer is the same as adding a negative integer, which gives you a lower score.

Copyright © Big Ideas Learning, LLC
All rights reserved.

Answers for Resources by Chapter

7. *Sample answer:* Students may use a strategy such as aiming over 7 for the first drop and over -7 for the second drop, or if they notice their paper punches always fall in some direction, they might try to get the pieces to fall over values that will give them positive points.

11.3 Puzzle Time

NOTHING IT JUST WAVED

11.4 Start Thinking!

For use before Activity 11.4

Sample answer: A deposit is adding a positive amount of money to your account. A withdrawal is adding a negative amount to your account.

11.4 Warm Up

For use before Activity 11.4

1. 18 2. −14 3. −9
4. 15 5. 24 6. −16

11.4 Start Thinking!

For use before Lesson 11.4

Sample answer: A problem like $2(4)$ means to add 2 groups of 4.

11.4 Warm Up

For use before Lesson 11.4

1. 14 2. −63 3. −48
4. 80 5. −30 6. 60

11.4 Practice A

1. −12 2. −30 3. 16 4. 54
5. 0 6. 36 7. 77 8. −25
9. −91 10. −9 11. −24 12. 81
13. −35 14. −30 15. −20 16. −42
17. 0 18. 48 19. 100 20. 9
21. −9 22. −8 23. −25 24. −48
25. 98 26. −18 27. 22 28. −7
29. 96, −192 30. 729, −2187

31. a. −12 ft

b.

Time	3 sec	6 sec	9 sec
Height	144 ft	108 ft	72 ft

c. 15 sec d. −48 ft

11.4 Practice B

1. 96 2. −140 3. −84 4. 120
5. −384 6. 220 7. −270 8. 0
9. −156 10. 96 11. −729 12. 144
13. −144 14. −343 15. 8 16. −72
17. 847 18. 18 19. −19 20. 24

21. a. $325

b. Because $n = 0$ represents the 1st member, the 8th member pays $25. So, the 9th member is free.

c. $25, the 8th member

22. a. 47 b. yes; $-3, 16$ c. -49

11.4 Enrichment and Extension

1. 1575 2. 432 3. −11,340 4. −2000

5.

$$(-3)^2 = (-3)(-3) = 9$$
$$(-3)^3 = (-3)(-3)(-3) = -27$$
$$(-3)^4 = (-3)(-3)(-3)(-3) = 81$$
$$(-3)^5 = (-3)(-3)(-3)(-3)(-3) = -243$$
$$(-3)^6 = (-3)(-3)(-3)(-3)(-3)(-3) = 729$$
$$(-3)^7 = (-3)(-3)(-3)(-3)(-3)(-3)(-3) = -2187$$
$$(-3)^8 = (-3)(-3)(-3)(-3)(-3)(-3)(-3)(-3) = 6561$$

6. The sign changes every time.

7. positive; The products are positive when the exponent is even.

8. negative; The products are negative when the exponent is odd.

11.4 Puzzle Time

DURING LEAP YEAR

11.5 Start Thinking!

For use before Activity 11.5

$+ \times +,\ + \times -,\ - \times +,\ - \times -$

11.5 Warm Up

For use before Activity 11.5

1. −50 2. 30 3. −72
4. −105 5. 176 6. −128

Copyright © Big Ideas Learning, LLC
All rights reserved.

Answers for Resources by Chapter

11.5 Start Thinking!

For use before Lesson 11.5

24, 96; Divide by 2; Multiply by 4

11.5 Warm Up

For use before Lesson 11.5

1. -4 **2.** -8 **3.** 7

4. -5 **5.** -9 **6.** 12

11.5 Practice A

1. -2 **2.** 5 **3.** -2 **4.** 0

5. -5 **6.** -3 **7.** -8 **8.** -1

9. undefined **10.** -6

11. 7 **12.** -7 **13.** 4 **14.** 3

15. -5 **16.** 10 **17.** -12 **18.** 6

19. -6, 3 **20.** 20, -4 **21.** -7

22. a. -4 **b.** 15, down **c.** down

d. Between 4 seconds and 5 seconds because the sign changed.

11.5 Practice B

1. -17 **2.** -3 **3.** 0 **4.** 16

5. -21 **6.** undefined

7. -12 **8.** 1 **9.** -11 **10.** -9

11. -52 **12.** 16 **13.** 3 **14.** -11

15. 2 **16.** 28 **17.** 50 **18.** 12

19. a. 2 **b.** 2

c. *Sample answer:* PI-Squared, because they answer both parts of more questions.

20. a. 7 hours **b.** 35 hours **c.** 8 hours 10 minutes

11.5 Enrichment and Extension

1. sometimes **2.** sometimes **3.** never

4. always **5.** always **6.** never

7. never **8.** always **9.** sometimes

10. sometimes **11.** always **12.** never

13. never **14.** always **15.** always

16. never **17.** none **18.** all integers

19. all positive integers and zero

20. all integers

21. all positive integers and zero

22. all negative integers and zero

23. all negative integers and zero

24. all positive integers and zero

25. all positive integers and zero

26. all negative integers and zero

27. all integers

28. all integers

29. all positive integers and zero

30. all positive integers and zero

31. all negative integers and zero

32. all negative integers and zero

Check students' work; It should be a person.

11.5 Puzzle Time

CATCH YOU LATER

Technology Connection

1. -8 **2.** 0 **3.** -1

4. -1 **5.** -7 **6.** -2

7. *Sample answer:* The number becomes 1 smaller, meaning it is 1 farther away from the zero. The magnitude of the number—its absolute value—becomes 1 larger, again because it moves 1 unit farther from zero.

Chapter 12

12.1 Start Thinking!

For use before Activity 12.1

you did; *Sample answer:* If you change the fractions to have common denominators, $\frac{3}{8}$ is greater than $\frac{2}{8}$ or $\frac{1}{4}$.

12.1 Warm Up

For use before Activity 12.1

1. $=$ **2.** $<$ **3.** $>$

4. $>$ **5.** $<$ **6.** $=$

Copyright © Big Ideas Learning, LLC
All rights reserved.

Answers for Resources by Chapter

12.1 Start Thinking!

For use before Lesson 12.1

The first friend is the better shooter. *Sample answer:* First write each shooter's score as a ratio (rational number). Then change each to a decimal. Finally compare the decimals to see which is the largest.

12.1 Warm Up

For use before Lesson 12.1

1. $-\frac{7}{4}, -\frac{2}{3}, -0.3, 0.6, \frac{3}{4}$ **2.** $-1.3, -\frac{6}{5}, \frac{7}{5}, 1.5, 1.65$

3. $-2.75, -0.37, \frac{5}{4}, 2.65, \frac{11}{4}$

4. $-\frac{4}{2}, -0.8, \frac{1}{8}, \frac{4}{10}, 4.5$ **5.** $-\frac{9}{3}, -0.3, \frac{6}{4}, \frac{8}{5}, 3.8$

6. $-1.5, -\frac{3}{4}, 0.6, \frac{9}{6}, \frac{7}{3}$

12.1 Practice A

1. $0.\overline{5}$ **2.** -0.375 **3.** $-0.\overline{27}$ **4.** $0.2\overline{3}$

5. $1.41\overline{6}$ **6.** $-2.\overline{3}$ **7.** $-0.\overline{590}$ **8.** $5.1\overline{6}$

9. $\frac{7}{10}$ **10.** $-\frac{3}{10}$ **11.** $-\frac{43}{100}$ **12.** $\frac{13}{25}$

13. $1\frac{1}{4}$ **14.** $-2\frac{7}{100}$ **15.** $4\frac{9}{50}$ **16.** $3\frac{1}{8}$

17. $-\frac{5}{2}, -\frac{2}{3}, -0.5, \frac{3}{2}, 1.6$

18. $-\frac{7}{4}, -1.7, 0.6, \frac{3}{4}, 1.1$

19. $-0.5, -\frac{2}{5}, 0, 0.67, \frac{7}{9}$ **20.** $-\frac{1}{3}, -0.3, 1.2, \frac{4}{3}, \frac{3}{2}$

21. a. 0.64 **b.** $\frac{16}{25}$ **22.** Ian

23. a. $77.\overline{27}\%$ **b.** yes **c.** 16

12.1 Practice B

1. 0.625 **2.** $-0.1\overline{36}$ **3.** $1.\overline{2}$ **4.** -5.075

5. $-7.\overline{45}$ **6.** $4.0\overline{6}$ **7.** $-9.\overline{1}$ **8.** $-7.8\overline{3}$

9. $\frac{17}{25}$ **10.** $-\frac{1}{100}$ **11.** $-3\frac{99}{100}$ **12.** $8\frac{149}{200}$

13. $3\frac{1}{200}$ **14.** $-13\frac{3}{250}$

15. $-9\frac{49}{50}$ **16.** $-10\frac{113}{250}$

17. your friend **18.** >

19. > **20.** <

21. *Sample answer:* $-1.75, -1.\overline{6}$

22. a. $-2\frac{6}{7}, -\frac{3}{2}, 2.25, 2\frac{1}{3}$

b. 10:00 A.M. and 4:00 P.M.

c. $-2\frac{6}{7}$ feet **d.** decrease

e. increase; The tide increased at 4:00 A.M. on the given day, so it will increase again in the morning.

12.1 Enrichment and Extension

1. *Sample answer:* There are positive and negative numbers in the outer box. The larger circle has just negative integers and the smaller circle has just positive integers.

2.

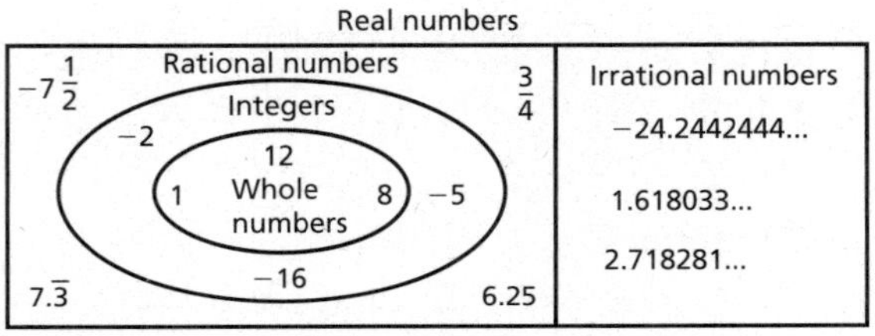

3. rational **4.** rational **5.** irrational

6. rational, integer, whole number

7. yes; $2 = \frac{2}{1}$; All whole numbers are rational numbers because they can be written as a fraction with a denominator of 1.

8. yes; $-14 = \frac{-14}{1}$; All integers are rational numbers because they can be written as a fraction with a denominator of 1.

9. sometimes **10.** always **11.** never

12. sometimes **13.** always

12.1 Puzzle Time

THE STUDENT WHO SAID HE COULDN'T WRITE HIS ESSAY ON GOLDFISH BECAUSE HE DIDN'T HAVE ANY WATERPROOF INK

Copyright © Big Ideas Learning, LLC
All rights reserved.

Answers for Resources by Chapter

12.2 Start Thinking!

For use before Activity 12.2

Sample answer: Add $55 + 15$, then add the answer to -20, $70 + (-20)$. Finally add the answer from the previous problem to 8, $50 + 8$. The final answer $58°$. So, the temperature at the end of the day is greater than the temperature at the beginning of the day.

12.2 Warm Up

For use before Activity 12.2

1. -77 **2.** 54 **3.** 42

4. -20 **5.** 33 **6.** -99

12.2 Start Thinking!

For use before Lesson 12.2

Sample answer: Start at zero and move $\frac{4}{5}$ unit to the right. Then move $\frac{2}{5}$ unit to the left and end on $\frac{2}{5}$.

Start at zero and move 2.6 units to the left. Then move 5.8 units to the right and end at 3.2.

12.2 Warm Up

For use before Lesson 12.2

1. $2\frac{7}{9}$ **2.** $-2\frac{4}{9}$ **3.** $-\frac{11}{15}$

4. -6.9 **5.** -18.2 **6.** -2.5

12.2 Practice A

1. $-\frac{1}{8}$ **2.** $\frac{1}{3}$ **3.** $\frac{1}{6}$

4. 4.3 **5.** -2.8 **6.** -5.61

7. The second number is negative.

$$\frac{3}{10} + \left(-\frac{1}{10}\right) = \frac{3 + (-1)}{10} = \frac{2}{10} = \frac{1}{5}$$

8. $-\frac{9}{10}$ **9.** $-\frac{3}{10}$ **10.** $\frac{1}{10}$ **11.** -4.7

12. $\frac{5}{8}$ **13.** $-\frac{1}{2}$ **14.** $-1\frac{14}{15}$ **15.** 2.4

16. a. always **b.** sometimes

c. never **d.** sometimes

12.2 Practice B

1. $\frac{1}{5}$ **2.** $-1\frac{1}{12}$ **3.** $-3\frac{1}{14}$

4. -6.17 **5.** -2.34 **6.** -28.51

7. $2\frac{5}{6} + \left(-\frac{8}{15}\right) = \frac{17}{6} + \left(-\frac{8}{15}\right) = \frac{85 + (-16)}{30}$

$= \frac{69}{30} = 2\frac{9}{30} = 2\frac{3}{10}$

8. $-\frac{19}{20}$ **9.** $-\frac{1}{20}$ **10.** $\frac{11}{20}$ **11.** $8.44

12. $\frac{13}{21}$ **13.** $-5\frac{11}{20}$ **14.** $-3\frac{7}{9}$ **15.** 20

16. When the positive number is greater than the absolute value of the negative number.

17. The snowfall for the three month period is $2\frac{1}{3}$ inches greater than the yearly average.

18. $-$2.35

12.2 Enrichment and Extension

1. $\boxed{3}\frac{\boxed{2}}{\boxed{5}} - 4\frac{\boxed{1}}{2} = -1\frac{1}{10}$

2. $-\boxed{2}.5\boxed{8} + \boxed{3}7.\boxed{9} = 35.32$

3. $\boxed{2}\boxed{7}.\boxed{8} - 4\boxed{3}8 = -410.2$

4. $-\frac{\boxed{3}}{\boxed{4}} - \boxed{2}\frac{\boxed{1}}{2} = -3\frac{1}{4}$

5. $-\boxed{3}.8\boxed{9} - \boxed{4}2.\boxed{3} = -46.19$

6. $\boxed{6}\frac{\boxed{7}}{8} - \boxed{8}\frac{3}{\boxed{4}} = -1\frac{7}{8}$

7. $-\boxed{6}\frac{\boxed{2}}{3} + 2\frac{\boxed{1}}{\boxed{9}} = -4\frac{5}{9}$

8. $-\boxed{3}8 + \boxed{5}\boxed{6}.\boxed{4} = 18.4$

9. *Sample answer:*

$\boxed{5}.\boxed{6}\boxed{1} - \boxed{7}.\boxed{4}\boxed{3} = -1.\boxed{8}\boxed{2}$

12.2 Puzzle Time

AT THE NORTH POLE

Copyright © Big Ideas Learning, LLC
All rights reserved.

Answers for Resources by Chapter

12.3 Start Thinking!

For use before Activity 12.3

Sample answer. One example of a sport is swimming. Each leg of a relay team is added together to get a final time. A way negative rational numbers is used in swimming is when a swimmer's time is below a record time.

12.3 Warm Up

For use before Activity 12.3

1. −70 **2.** 91 **3.** −75

4. −75 **5.** 12 **6.** 42

12.3 Start Thinking!

For use before Lesson 12.3

Sample answer: Start at zero and move $\frac{4}{5}$ unit to the left. Then move $\frac{2}{5}$ unit more to the left and end on $-1\frac{1}{5}$.

Start at zero and move 2.6 units to the left. Then move 5.8 units more to the left and end at −8.4.

12.3 Warm Up

For use before Lesson 12.3

1. $-7\frac{5}{7}$ **2.** $-13\frac{1}{2}$ **3.** $-10\frac{5}{8}$

4. −18.9 **5.** 23.8 **6.** −18

12.3 Practice A

1. −1 **2.** $1\frac{2}{3}$ **3.** $1\frac{11}{12}$ **4.** $-5\frac{1}{3}$

5. −2.46 **6.** −6.25 **7.** $10\frac{1}{4}$ **8.** 2.6

9. $3\frac{1}{15}$ **10.** $\frac{5}{12}$ **11.** $-\frac{7}{8}$ **12.** −2.57

13. a. $3\frac{2}{3}$ gal **b.** $4\frac{11}{12}$ gal **c.** $1\frac{7}{12}$ gal

14. no; If the absolute value of the second number is greater than the absolute value of the first number, then the difference is negative.

12.3 Practice B

1. $1\frac{4}{5}$ **2.** $\frac{1}{3}$ **3.** $4\frac{1}{12}$

4. $-13\frac{17}{24}$ **5.** −42.481 **6.** 3.786

7. $2\frac{8}{15}$ **8.** 13.7 **9.** 1.7

10. $7\frac{13}{16} - 9\frac{5}{8} = -1\frac{13}{16}$

11. $-5\frac{5}{6}$ **12.** −24.625 **13.** 3.975

14. $-1\frac{2}{9}$ **15.** −\$90.73 **16.** $1\frac{6}{8}$

12.3 Enrichment and Extension

1. The golfer is 4 strokes above par.

2. There is \$3.26 left in the account.

3. The team lost 2 yards.

4. negative; The amount in the bank account is \$1.87 less than the cost of the clothing.

5. positive; Distance is always positive.

6. negative; $9 - 14 = -5$.

7. zero; $6 - 3\frac{2}{3} = 2\frac{1}{3}$ and $2\frac{1}{3} - 2\frac{1}{3} = 0$.

12.3 Puzzle Time

ON A BED OF LETTUCE

12.4 Start Thinking!

For use before Activity 12.4

Gain of 2 yards (+2); Loss of 8 yards (−8);
Gain of 20 yards (+20); Loss of 5 yards (−5);
Gain of 3 yards (+3)

12.4 Warm Up

For use before Activity 12.4

1. −108 **2.** −110 **3.** 168

4. 8 **5.** −14 **6.** −27

12.4 Start Thinking!

For use before Lesson 12.4

The company shows a loss of \$2.23.

12.4 Warm Up

For use before Lesson 12.4

1. $-1\frac{4}{5}$ **2.** $-\frac{27}{125}$ **3.** 3.66

4. $\frac{6}{7}$ **5.** −1.2 **6.** −0.215

Copyright © Big Ideas Learning, LLC
All rights reserved.

Answers for Resources by Chapter

12.4 Practice A

1. negative 2. negative 3. positive 4. positive

5. $-\frac{1}{5}$ 6. $\frac{2}{3}$ 7. $-\frac{1}{21}$ 8. $-\frac{7}{10}$

9. 0.8 10. −1.5 11. $-\frac{4}{7}$ 12. $\frac{5}{6}$

13. $-3\frac{1}{3}$ 14. $1\frac{57}{64}$ 15. −7.77 16. 11.742

17. 10 pizzas 18. −2.45 in. 19. −9.46

20. $\frac{6}{11}$ 21. −14.2979 22. $2\frac{1}{2}$

23. *Sample answer:* $-\frac{5}{4}, -\frac{1}{2}$

24. −3.4

12.4 Practice B

1. $-1\frac{4}{11}$ 2. $\frac{10}{39}$ 3. $-\frac{11}{180}$ 4. $-\frac{21}{80}$

5. 8.9 6. −6.3 7. $\frac{5}{9}$ 8. $-\frac{3}{8}$

9. $-4\frac{5}{6}$ 10. $-11\frac{14}{25}$ 11. −8.525 12. 2.5654

13. 16 burgers 14. −0.52 second

15. 0.241 16. $3\frac{1}{2}$

17. −28.5939 18. $-11\frac{9}{25}$

19. a. $46.25 b. $72.52

12.4 Enrichment and Extension

1. 24 Zweenubs, 16 Zweedulls, 6 Zweebuds

2. $218.32

3. 20 Zweenubs, 10 Zweedulls, 2 Zweebuds

4. $198.88 5. −$19.44 6. −$97.20

7. 25 Zweenubs, 10 Zweedulls, 5 Zweebuds

8. They need to make at least $64.21. The most profit they could earn is $78.70.

9. No, they cannot make enough income on Zweenubs alone. Because only 25 Zweenubs are being manufactured next week, and last week's cannot be reused, there are only 13 Zweenubs left to sell. When you multiply 13 by $4.50, you get a total possible income of $58.50 from Zweenubs, which is less than the $64.21 that is needed in order to break even.

12.4 Puzzle Time

WHEN HE DRIBBLES

Technology Connection

1. $\frac{9}{10}$ 2. $2\frac{3}{8}, \frac{19}{8}$ 3. $\frac{-13}{36}$

4. $3\frac{7}{75}, \frac{232}{75}$ 5. $\frac{13}{16}$ 6. $1\frac{11}{36}, \frac{47}{36}$

Chapter 13

13.1 Start Thinking!

For use before Activity 13.1

Sample answer: An expression can contain numbers, variables, and operations (i.e. $x + 9$, $3 + 4$). An equation can contain numbers, variables, operations, and has an equal sign (i.e. $x + 3 = 8$, $1 + 2 = 3$).

13.1 Warm Up

For use before Activity 13.1

1. $7 + x$ 2. $-14 - y$ 3. $-19 + n$

4. $14y$ 5. $\frac{10}{n + 6}$ 6. $6\left(\frac{x}{3}\right)$

13.1 Start Thinking!

For use before Lesson 13.1

$4x + 4$; *Sample answer:* Your brother incorrectly combined the $6x$ and the $2x$. He should have subtracted instead of added.

13.1 Warm Up

For use before Lesson 13.1

1. $14x$ 2. $7y - 3$ 3. $2x + 16$

4. $11y - 4$ 5. $10x - 15.4$ 6. $y + 5\frac{1}{2}$

13.1 Practice A

1. $-4y, 7, 9y, -3$; $-4y$ and $9y$, 7 and -3

2. $3n^2, -1.4n, 5n^2, -6.4$; $3n^2$ and $5n^2$

Copyright © Big Ideas Learning, LLC
All rights reserved.

Answers for Resources by Chapter

3. $\frac{1}{2}b^3, -b^3, 2b;\ \frac{1}{2}b^3$ and $-b^3$

4. $-6m$ **5.** $14k - 8$ **6.** $10.3 - 12x$

7. $13 - 8x$ **8.** $28a - 7$ **9.** $27x - \frac{89}{6}$

10. $-\frac{5}{8}h + 7$ **11.** $-\frac{1}{4}y + 2$ **12.** $18x + 16$

13. $8y + 2.1y + 3y$

The total weight carried by the runners is $8y + 2.1y + 3y = 13.1y$ ounces.

14. $65 + 35x$ kilograms **15.** yes; $15a^2 - 4b$

13.1 Practice B

1. $1.3x, -2.7x^2, -5.4x, 3;\ 1.3x$ and $-5.4x$

2. $10, -\frac{3}{10}m, 6m^2, \frac{2}{5}m;\ -\frac{3}{10}m$ and $\frac{2}{5}m$

3. $-\frac{35}{12}b$ **4.** $180m - 40$ **5.** $31.4 - 14x$

6. $-12y + 12$ **7.** $-7v - 3$ **8.** $\frac{17}{15}x + \frac{79}{5}$

9. $29x + 9$

10.

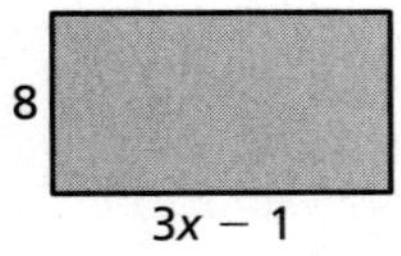

11.

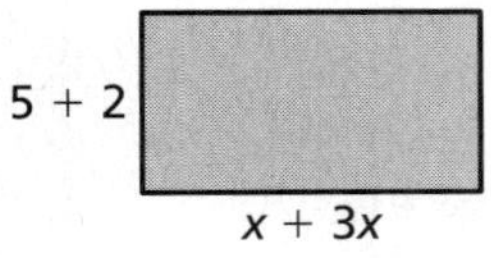

12. $2.5x + 5$ **13.** $6w + 60$ **14.** $15.65x$

13.1 Enrichment and Extension

1. a **2.** c **3.** j **4.** g

5. d **6.** i **7.** f **8.** h

9. k **10.** b **11.** e

12. Answers will vary. **13.** Answers will vary.

13.1 Puzzle Time

THROW IT UP IN THE AIR. IT WILL COME DOWN AND SQUASH.

13.2 Start Thinking!

For use before Activity 13.2

Sample answer: like terms: $3x$, $24x$; unlike terms: $9y$, $9m$; Like terms are terms that have the same variable raised to the same exponent. Unlike terms are terms that cannot be combined further.

13.2 Warm Up

For use before Activity 13.2

1. $x + 2$ **2.** $-2y - 9$ **3.** $-10x + 3$

4. $3\frac{1}{2}y - \frac{2}{5}$ **5.** $8.7x - 4.6$ **6.** $5.1y - 3.3$

13.2 Start Thinking!

For use before Lesson 13.2

Model the expression using algebra tiles.

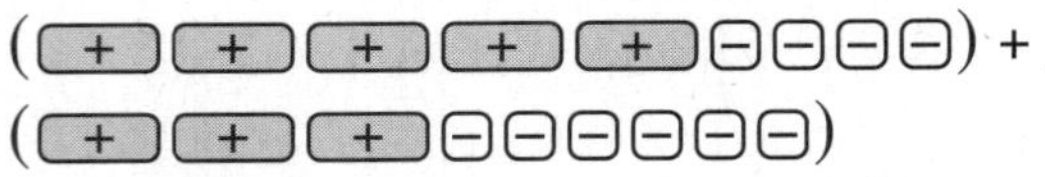

Then combine the like terms. The simplified expression is $8x - 10$.

13.2 Warm Up

For use before Lesson 13.2

1. $(x - 4) + (x + 2);\ 2x - 2$

2. $(2x - 3) + (x - 3);\ 3x - 6$

3. $(x + 2) - (2x + 3);\ -x - 1$

4. $(2x - 4) - (x + 2);\ x - 6$

13.2 Practice A

1. $2p - 10$ **2.** $2n + 3$

3. $2r + 7$ **4.** $8x - 27$

5. $8.2c - 18$ **6.** $28.5q - 2$

7. $6.5y + 13$ **8.** $6x - 3$

9. a. $15t + 2$ **b.** your friend

10. $-2k + 8$ **11.** $-8d - 5$ **12.** $19j - 9$

13. $-12x + 26$ **14.** $5t - 18$ **15.** $4w + 5$

16. a. $4 + 0.5r$ **b.** \$7.50

17. $10m + 2$

Copyright © Big Ideas Learning, LLC
All rights reserved.

Answers for Resources by Chapter

13.2 Practice B

1. $2t + 7$
2. $11k + 6$
3. $-8y + 4$
4. $13.6g - 22$
5. $-8.3s$
6. $3p + 1$
7. $-5w - \frac{5}{2}$
8. $4k - \frac{13}{5}$
9. a. $5t + 7$ b. $29t + 45$ c. \$132
10. $-6u + 7$
11. $-28x + 55$
12. $11.4h - 26$
13. $4.5b + 24$
14. $2j - 8$
15. $6n + \frac{22}{5}$
16. a. $7d + 12$ b. 5 pairs

13.2 Enrichment and Extension

1. $x^3 + x^2$
2. $-2x^2 + 8x$
3. $x^5 - 4x$
4. $3x^2 - x$
5. $3x^2 - 3x$
6. $2x^2 - 2x$
7. $-16x^2 - 12x$
8. $n^2 - 4n$
9. $-3b^2 - 9b$
10. $-8w^2 - 28w$
11. $-4x^2 - 12x$
12. $-15k^2 + 68k + 18$
13. $2.2h^2 + 12.2h + 4$
14. $m^2 + 3m$
15. $-z^2 - 2z - 2$
16. $96d^9 + 18d^6$

13.2 Puzzle Time

YOU LIGHT UP MY LIFE

Extension 13.2 Start Thinking!

For use before Extension 13.2

no; The common factors of 15 and 30 are 1, 3, 5, and 15. So, the greatest common factor is 15.

Extension 13.2 Warm Up

For use before Extension 13.2

1. 2
2. 3
3. 9
4. 7
5. 6
6. 10

Extension 13.2 Practice

1. $2(4 - 11)$
2. $5(5 + 6)$
3. $3(2y + 1)$
4. $2(t - 5)$
5. $8(2p - 1)$
6. $3(7s + 5)$
7. $8(4v + 3w)$
8. $3(3b + 8c)$
9. $6(2y - 7z)$
10. $\frac{1}{2}(m + 1)$
11. $\frac{2}{3}\left(j - \frac{1}{3}\right)$
12. $1.2(k + 2)$
13. $1.5(a - 3)$
14. $3\left(f + \frac{5}{3}\right)$
15. $\frac{3}{10}(x - 2)$
16. $-\frac{1}{3}(x + 36)$
17. $-\frac{1}{6}(2x - 5y)$
18. $(3x - 2)$ in.
19. a. $6x + 9$ b. $21x + 28$ c. $15x + 19$

13.3 Start Thinking!

For use before Activity 13.3

$p - 7$ and $p + 8$; The first expression can only be written one way. The second expression could be written as $8 + p$ because addition is commutative.

13.3 Warm Up

For use before Activity 13.3

1. 42
2. −46
3. 7
4. −39
5. −6
6. 65

13.3 Start Thinking!

For use before Lesson 13.3

Sample answer: In the equation $p + 4 = 10$, subtraction will be the inverse operation used to solve this problem. By subtracting 4 from each side, the variable p will be left alone on the left side resulting in the answer 6.

13.3 Warm Up

For use before Lesson 13.3

1. $x = 5$
2. $y = 18$
3. $n = 78$
4. $p = 15$
5. $t = 57$
6. $z = 7$

13.3 Practice A

1. 7
2. −8
3. −4
4. 9.6
5. −10.7
6. $1\frac{1}{6}$
7. −18.82
8. $3\frac{5}{8}$
9. 3.94
10. $-2\frac{5}{12}$
11. $\frac{11}{14}$
12. 2.06
13. $y + 5 = -2; -7$
14. $8 + h = 12; 4$

Copyright © Big Ideas Learning, LLC
All rights reserved.

15. $-13 = n - 4; -9$ **16.** $b + 2 = 9; \$7$

17. $325 = t - 75; 400°F$

18. $c - 29\frac{3}{4} = -10\frac{1}{4}; 19\frac{1}{2}$ in.

19. $76.50 = f - 31.41; \$107.91$

20. $5.2 + 8 + x = 25; 11.8$ ft

21. -4

13.3 Practice B

1. 88 **2.** -36 **3.** -6.4 **4.** 19.63

5. -7.143 **6.** 7 **7.** -109.04 **8.** $-1\frac{1}{2}$

9. 4.994 **10.** $-9\frac{11}{24}$ **11.** $6\frac{11}{15}$ **12.** -1.798

13. $27 = x + 12; 15$ **14.** $p - (-9) = 12; 3$

15. $m - 35 = -72; -37$

16. $28.12 = f - 0.14$; 28.26 seconds

17. $7\frac{3}{8} + \ell = 16\frac{1}{4}; 8\frac{7}{8}$ meters

18. $t - 183.6 = -109.3; 74.3°F$

19. \$53.30 **20.** $14.8, -14.8$ **21.** $10, -14$

13.3 Enrichment and Extension

1. yes

2. Karen's answer is more complete because she remembered that –7 has an absolute value of 7.

3. *Sample answer:* Plot 12 on a number line and then find the numbers that are 8 units away from 12.

4. yes; *Sample answer:* Because the absolute value of the expression $x - 5$ equals 9, then the expression $x - 5$ is equal to either 9 or -9. So, set up two equations and solve to find x.

5. Absolute value cannot be negative, but in this case x is inside the absolute value bars. So, x can be negative because you can take the absolute value of a negative number.

6. *Sample answer:* He substituted -5 for $|x|$, and the absolute value of a number is never negative.

7. Every absolute value equation does not have two solutions. In Exercise 6, by subtracting 7 from each side, you obtain $|x| = -5$. This equation has no solutions because the absolute value of a number is never negative.

8. a. $|x| + 6 = 6$ **b.** $|x + 3| = 5$

c. $|x| + 8 = 4$

13.3 Puzzle Time

LOOK MOM NO HANDS

13.4 Start Thinking!

For use before Activity 13.4

Sample answer: To determine if you have enough money to buy the items, you will need to find the total price. You can do this by adding the price together, then comparing the total with how much you have. If you are buying more than one item of the same price, you can use multiplication to find the price.

13.4 Warm Up

For use before Activity 13.4

1. -168 **2.** 216 **3.** -65

4. -32 **5.** -9 **6.** 16

13.4 Start Thinking!

For use before Lesson 13.4

Sample answer: A small bus has 13 rows. Write and solve an equation to find the number of students each seat holds if there are a total of 39 students riding the bus. A small bus carries a total of 39 students. Each seat holds 3 students. Write and solve an equation showing how many rows of seats are on the bus.

13.4 Warm Up

For use before Lesson 13.4

1. $x = 6$ **2.** $x = 28$ **3.** $x = -11$

4. $x = 60$ **5.** $x = 63$ **6.** $x = -48$

13.4 Practice A

1. 6 **2.** -5 **3.** -99 **4.** -160

5. 1.4 **6.** -5.6 **7.** -15 **8.** $4\frac{2}{3}$

9. -14.28 **10.** $-3\frac{1}{3}$ **11.** 3 **12.** -16

13. $\frac{1}{2}x = -\frac{5}{12}; -\frac{5}{6}$ **14.** $\frac{n}{0.2} = -2.6; -0.52$

Copyright © Big Ideas Learning, LLC
All rights reserved.

15. $7.50h = 123.75$; 16.5 hours

16. $\frac{m}{4.5} = 70$; 315 miles

17. $\frac{3}{8}w = \frac{1}{2}$; $1\frac{1}{3}$ in.

18. $\frac{s}{3} = 21$; 63 students

19. $26.46 = 4s$; 6.615 in.

20. *Sample answer:* $7n = 2$

21. *Sample answer:* $\frac{y}{2} = -10$

13.4 Practice B

1. 3.75 **2.** 1.5 **3.** −36.75 **4.** −3.96

5. 7.5 **6.** −2.35 **7.** $-4\frac{4}{5}$ **8.** $0.\overline{45}$

9. −0.0084 **10.** $1\frac{1}{8}$ **11.** 9 **12.** −33

13. $12t = 0.78$; 0.065, or 6.5%

14. $72 = \frac{8}{9}p$; 9 points

15. *Sample answer:* $3n = -44.4$

16. *Sample answer:* $\frac{m}{9} = -\frac{1}{14}$

17. There are 5 students in each group and 2 students are not in a group.

18. a. $1.75y = 15.75$; 9 tokens

b. $12t = 17.50$; about \$1.46

c. $35 = 1.40t$; 25 free tokens

19. 6, −6

13.4 Enrichment and Extension

1. one solution; The value of x is positive because a negative must be multiplied by a positive when the result is negative.

2. one solution; The value of n is positive because a positive must be divided by a negative when the result is negative.

3. one solution; The value of g is negative because the number divided by a positive must be negative when the result is negative.

4. one solution; The value of t is negative because a negative must be multiplied by a negative when the result is positive.

5. two solutions; The value of v is either positive or negative, because the absolute value will always be positive, and a negative must be multiplied by a positive when the result is negative.

6. no solution; The absolute value of k will always be positive, and a positive divided by a negative cannot be equal to a positive.

7. no solution; The absolute value of the quotient of x and -5 will always be positive and cannot be equal to -15.

8. two solutions; The value of p is positive or negative, because the absolute value of the product of -6 and p will always be positive.

9. two solutions; The value of u is positive or negative, because the absolute value of the product of 2.7 and u will always be positive.

10. no solution; The absolute value of the product of $1\frac{1}{2}$ and b will always be positive and cannot be equal to $-13\frac{4}{5}$.

11. one solution; The value of h is positive, because the absolute value of 8 is positive, and a positive must be multiplied by a positive when the result is positive.

12. one solution; The value of y is negative, because the absolute value of -9 is positive, and a positive must be multiplied by a negative when the result is negative.

13. two solutions; The value of b is either positive or negative, because the absolute value will always be positive, and the number divided by a negative must be positive when the result is negative.

14. no solution; The absolute value of the quotient of a and -5 will always be positive and cannot be equal to -2.5.

Check students' work; It should be a fish.

13.4 Puzzle Time

THE RESTAURANT ON THE MOON THAT HAS REALLY GOOD FOOD BUT NO ATMOSPHERE

Copyright © Big Ideas Learning, LLC
All rights reserved.

Answers for Resources by Chapter

13.5 Start Thinking!

For use before Activity 13.5

Sample answer: $x + 2 = 4$; To solve this equation subtract 2 from each side of the equation. The answer is $x = 2$.

$x - 3 = 5$; To solve this equation add 3 to each side of the equation. The answer is $x = 8$.

$4x = 16$; To solve this equation divide each side of the equation by 4. The answer is $x = 4$.

$\frac{x}{5} = 3$; To solve this equation multiply each side of the equation by 5. The answer is $x = 15$.

13.5 Warm Up

For use before Activity 13.5

1. $x = 12$ **2.** $x = -9$ **3.** $x = -40$

4. $x = 12$ **5.** $x = 9.04$ **6.** $x = 4\frac{1}{6}$

13.5 Start Thinking!

For use before Lesson 13.5

Add 4 to both sides of the equation. Divide both sides by 3. The solution is $x = 5$.

13.5 Warm Up

For use before Lesson 13.5

1. $x = 8$ **2.** $x = -12$ **3.** $x = 5$

4. $x = -6$ **5.** $x = -2.65$ **6.** $x = -5.2$

13.5 Practice A

1. 4 **2.** −2.4 **3.** 3.5 **4.** 4.4

5. −3 **6.** 7.6 **7.** −1.05 **8.** −1.5

9. 17.35 **10.** −1 **11.** $-\frac{1}{2}$ **12.** −4

13. $2.5r + 4 = 21.50$; 7 rides

14. $45 + 1.99m = 68.88$; 12 movies

15. $16 + 2w = 24$; 4 feet

16. 9 **17.** $-\frac{1}{3}$ **18.** −6

19. 32 students

20. a. 10 cans **b.** $x = 10$; same answer

13.5 Practice B

1. 3 **2.** −4 **3.** 0.78125 **4.** −7

5. $2.\overline{8}$ **6.** −1.32 **7.** $-2\frac{5}{14}$ **8.** 0

9. $2.708\overline{3}$ **10.** $\frac{5}{48}$ **11.** 43 **12.** $\frac{13}{15}$

13. $132.49 + 15s = 192.49$; 4 skateboards

14. $f + 27(0.99) = 42.72$; \$15.99

15. −2.25 **16.** −9.4 **17.** $6\frac{1}{6}$

18. 16 ft, 32 ft

19. a. 56 seashells **b.** 297 seashells

c. *Sample answer:* 50, 3

13.5 Enrichment and Extension

1. 10

2. Any multiple of 10 will work because 2 and 5 will be factors of it.

3. The number 10 because it is the least common denominator of the fractions.

4. Because of the Multiplication Property of Equality

5. Multiply each term of the equation by the least common denominator of the fractions.

6. a. 46 **b.** 15 **c.** 6 **d.** $1\frac{6}{7}$ **e.** $-1\frac{2}{9}$ **f.** −6

13.5 Puzzle Time

DON'T STOP ME I'M ON A ROLL

Technology Connection

1. 47 **2.** 1.28 **3.** 2618 **4.** 5668

5. 1593 **6.** 938 **7.** 81 **8.** 3.7

9. 305.2 **10.** 5.135

Chapter 14

14.1 Start Thinking!

For use before Activity 14.1

Sample answer: For flying: You would need to consider the cost of the ticket and parking at the airport. For driving: You would need to consider the cost of gas and travel time.

Copyright © Big Ideas Learning, LLC
All rights reserved.

Answers for Resources by Chapter

14.1 Warm Up

For use before Activity 14.1

1. 0.5 h **2.** 240 min **3.** 0.25 min

4. 2.5 days **5.** 72 h **6.** 168 h

14.1 Start Thinking!

For use before Lesson 14.1

your sister; *Sample answer:* Her erasers cost \$0.31 each, and your erasers cost \$0.33 each.

14.1 Warm Up

For use before Lesson 14.1

1. \$42 **2.** \$48 **3.** 200 mi

4. \$5850 **5.** \$30 **6.** 210 in.

14.1 Practice A

1. \$60 **2.** \$3.54 **3.** 630 **4.** $\frac{4}{5}$

5. $\frac{8}{3}$ **6.** $\frac{7}{3}$ **7.** 60 mi/h

8. 3 bowlers per lane **9.** \$4 per person

10. $\frac{1 \text{ lap}}{2 \text{ min}}$, or 0.5 lap per min

11. 15 grams per serving **12.** 8 min/mi

13. yes; *Sample answer:*

$\frac{\text{blue}}{\text{orange}} = \frac{12}{9} = \frac{4}{3}$ and $\frac{\text{orange}}{\text{blue}} = \frac{3}{4}$

14. 26 students per classroom

15. **a.** *Sample answer:* 90-pack; Often the largest package has the lowest unit rate.

b. \$0.27 per pack

c. \$0.18 per pack

d. \$0.21 per pack

e. 60-pack; *Sample answer:* This is not what was expected.

14.1 Practice B

1. $\frac{5}{9}$ **2.** $\frac{3}{10}$ **3.** $\frac{9}{5}$

4. $\frac{500}{1}$ **5.** $\frac{29}{4}$ **6.** $\frac{3}{7}$

7. \$0.225 per can **8.** \$0.0645 per oz

9. about 1.82 m/sec **10.** 53 bacteria/h

11. **a.** B **b.** A **c.** B

d. *Sample answer:* protein **e.** calories

12. **a.** \$6 for 4 scoops **b.** \$1.50 per scoop **c.** \$18

d. *Sample answer:* The graph would be a line that would be above the given graph except at point (0, 0).

e. The point would become (4, 7).

13. *Sample answer:* It costs \$45 for 25 baseball tickets. What is the cost ratio?

14. *Sample answer:* You have 4 markers and 6 colored pencils. What is the ratio of markers to colored pencils?

14.1 Enrichment and Extension

1. \$15.60 **2.** \$30,784

3. Job A pays better. The hourly rate is \$15.60, which is higher than Job B's hourly rate of \$14.80. Also, Job A pays \$32,448 per year, when Job B only pays \$30,784 per year.

4. 180 miles **5.** 12.5 gallons **6.** 14.4 mi/gal

7. \$10 per day; \$50 per week; \$2600 per year

8. \$25

9. *Sample answer:* Sally should consider the length of time that the trips take to drive, the traffic for the different routes, road conditions, and whether or not there is public transportation available for all or part of the routes.

10. Job B's; Job A's macaroni and cheese contains 5.2 grams of fat per cup, and Job B's macaroni and cheese contains 5 grams of fat per cup.

11. Job A's; Job A's macaroni and cheese costs about \$0.45 per ounce, which is less expensive than the macaroni and cheese at the cafeteria of Job B.

12. *Sample answer:* Job A pays better, has the shorter and cheaper commute, and the cheaper macaroni and cheese. It is the best choice when you consider money.

13. *Sample answer:* It is important to consider whether or not you like the type of work that you will be doing. Also, it is important to enjoy the people you work with as well as the setting in which you are working. Also, the benefits, such as health insurance, that a job offers are very important.

14.1 Puzzle Time

A FIRE-QUACKER

Copyright © Big Ideas Learning, LLC
All rights reserved.

Answers for Resources by Chapter

14.2 Start Thinking!

For use before Activity 14.2

Sample answer: Two fractions that are equivalent are $\frac{15}{18}$ and $\frac{5}{6}$. To show that they are equivalent, reduce $\frac{15}{18}$ by dividing the numerator and denominator by 3. The reduced fraction equals $\frac{5}{6}$.

Two fractions that are not equivalent are $\frac{9}{14}$ and $\frac{13}{21}$. They are not equivalent because $\frac{9}{14}$ and $\frac{13}{21}$ are both in simplest form.

14.2 Warm Up

For use before Activity 14.2

1. $\frac{1}{2}$ **2.** $\frac{4}{3}$ **3.** $\frac{5}{4}$ **4.** $\frac{3}{7}$

5. $\frac{5}{7}$ **6.** $\frac{10}{41}$ **7.** 12 **8.** $\frac{10}{7}$

14.2 Start Thinking!

For use before Lesson 14.2

Sample answer: You determine which is the better buy by comparing how much cereal you get for the price.

14.2 Warm Up

For use before Lesson 14.2

1. yes **2.** no **3.** no **4.** yes

14.2 Practice A

1. yes **2.** yes **3.** no **4.** yes

5. no **6.** yes **7.** yes **8.** no

9. yes **10.** no **11.** yes **12.** yes

13. no

14. yes; *Sample answer:* Each unit rate is equivalent to 2.5 minutes per lap.

15. yes **16.** yes **17.** no

18. **a.** \$4.50 per chore

b. \$4.50 per chore

c. yes; The rates are equivalent.

19. no; 5 tickets for \$93.75 or 4 tickets for \$75.20

20. 1 potato and 3 carrots

14.2 Practice B

1. no **2.** yes **3.** yes **4.** yes

5. no **6.** no **7.** yes **8.** yes

9. no

10. yes; $\frac{15}{12} = \frac{5}{4}$ and $\frac{10}{8} = \frac{5}{4}$

11. no; 3 fluid ounces of vinegar

12. **a.** \$13 per pair

b. \$12 per pair

c. \$11.50 per pair

d. no; The cost per pair is different for each number of pairs of sandals that are purchased.

e. The buyer will purchase 30 pairs and get 3 free for \$379.50, purchase 5 pairs for \$65, and buy 2 pairs for \$30. The total cost is \$474.50.

13. 9 **14.** 15

14.2 Enrichment and Extension

1. Add 5 quarts of blue paint.

2. Add 1 quart of blue paint and 6 quarts of yellow paint.

3. Add 1 quart of yellow paint.

4. Add 1 pint of blue paint and 2 pints of red paint.

5. Add 2 pints of yellow paint and 1 pint of red paint.

6. Add 12 gallons of red paint.

7. Add 2 quarts of red paint and 4 quarts of blue paint.

8. Add 3 cups of yellow paint, 16 cups of red paint, and 4 cups of blue paint.

14.2 Puzzle Time

YOUR BREATH

Extension 14.2 Start Thinking!

For use before Extension 14.2

yes; Use the Cross Products Property to determine if the ratios are equal. Because $7 \times 6 = 42$ and $21 \times 2 = 42$, the ratios are equal.

Extension 14.2 Warm Up

For use before Extension 14.2

1. yes **2.** no

Copyright © Big Ideas Learning, LLC
All rights reserved.

Answers for Resources by Chapter

Extension 14.2 Practice

1. (0, 0): 0 pounds of chicken costs \$0.
(3, 12): 3 pounds of chicken costs \$12.
(5, 20): 5 pounds of chicken costs \$20.

2. (0, 0): You earn \$0 for working 0 hours.
(2, 50): You earn \$50 for working 2 hours.
(4, 100): You earn \$100 for working 4 hours.

3. $y = 2$ **4.** $y = 7$ **5.** $y = 6$ **6.** $y = 4$

7. a. Class A: 80%; Class B: 70% **b.** \$60 **c.** \$7.50

14.3 Start Thinking!

For use before Activity 14.3

Sample answer: The ratios $\frac{9}{2}$ and $\frac{27}{6}$ are proportional because they are equivalent fractions.

14.3 Warm Up

For use before Activity 14.3

1. *Sample answer:* $\frac{2}{3}, \frac{4}{6}$

2. *Sample answer:* $\frac{22}{40}, \frac{33}{60}$

3. *Sample answer:* $\frac{3}{7}, \frac{9}{21}$

4. *Sample answer:* $\frac{2}{5}, \frac{6}{15}$

5. *Sample answer:* $\frac{3}{5}, \frac{6}{10}$

6. *Sample answer:* $\frac{2}{5}, \frac{4}{10}$

7. *Sample answer:* $\frac{16}{18}, \frac{24}{27}$

8. *Sample answer:* $\frac{7}{12}, \frac{21}{36}$

14.3 Start Thinking!

For use before Lesson 14.3

Yes, the scores are proportional because they both scored 5 points in 1 minute.

14.3 Warm Up

For use before Lesson 14.3

1. $\frac{p}{100} = \frac{85}{100}$ **2.** $\frac{p}{50} = \frac{74}{100}$

3. $\frac{p}{25} = \frac{80}{100}$ **4.** $\frac{p}{110} = \frac{90}{100}$

14.3 Practice A

1. $\frac{p}{70} = \frac{90}{100}$ **2.** $\frac{p}{30} = \frac{72}{100}$

3. $\frac{f}{15} = \frac{60}{100}$ **4.** $\frac{f}{24} = \frac{75}{100}$

5. $\frac{2 \text{ hurricanes}}{6 \text{ storms}} = \frac{1 \text{ hurricane}}{n \text{ storms}}$

6. $\frac{w \text{ wins}}{21 \text{ races}} = \frac{8 \text{ wins}}{12 \text{ races}}$

7. $\frac{2 \text{ teachers}}{45 \text{ students}} = \frac{t \text{ teachers}}{315 \text{ students}}$

8. $a = 10$ **9.** $m = 77$ **10.** $d = 9$

11. a. $\frac{4}{9} = \frac{g}{36}$ **b.** 16 pints **c.** $6\frac{1}{2}$ gallons

12. a. 30 violin players **b.** 12 viola players **c.** $\frac{2}{5}$

13. *Sample answer:* $p = 4$, $q = 10$ and $p = 8$, $q = 20$

14.3 Practice B

1. $\frac{s}{32} = \frac{75}{100}$ **2.** $\frac{s}{80} = \frac{95}{100}$

3. The ratio of the length and height for each day should be set equal; $\frac{3.1}{h} = \frac{15.5}{45}$.

4. $\frac{3}{16} = \frac{r}{128}$ **5.** $x = 15$ **6.** $y = 136$

7. $r = 0.12$ **8.** $k = 120$ **9.** $p = 6.5$

10. $w = 32$ **11.** $\frac{1}{4}$ c

Copyright © Big Ideas Learning, LLC
All rights reserved.

12. a. $\frac{6}{7}, \frac{12}{14}, \frac{18}{21}$ **b.** $\frac{6}{11}, \frac{12}{22}$

c. 12 gray keys and 22 black keys

d. 12 gray keys, 22 black keys, 14 blue keys, 1 yellow key, 1 green key

14.3 Enrichment and Extension

1. 60 in.; 5 ft

2. *Sample answer:* If ramps are too steep, getting up and down a ramp in a wheelchair would be more difficult.

3. *Sample answer:* People in wheelchairs could get up and down more easily and be in less danger. However, the ramp will take up more room and cost more to build.

4. 30 ft

5. It meets the requirements. The ramp has a rise to run ratio of 1 : 14, which is less than 1 : 12.

6. The first sloping part is steeper; *Sample answer:* After a person pushes up the first part, the second part is not as steep and easier to push up and down when the person is more tired.

7. The first part must be 20 feet long and the second part must be 27.5 feet long.

8. *Answer should include, but is not limited to:* The design must show a maximum rise to run ratio of 1 : 12 with platforms at the top and bottom that are at least 60 inches long. Any changes in direction must have a platform that is at least 60 inches by 60 inches.

14.3 Puzzle Time

THE ORCA-DONTIST

14.4 Start Thinking!

For use before Activity 14.4

Sample answer: enlargements of pictures, doubling or halving recipes, mixing more paint

14.4 Warm Up

For use before Activity 14.4

1. 35 students per bus **2.** 121 miles per hour

3. 78 pages per hour **4.** 27 seats per row

5. 341 miles per hour **6.** 136 pages per day

14.4 Start Thinking!

For use before Lesson 14.4

Sample answer: The bakery will need to increase the amounts proportionally to what was made in the test batch.

14.4 Warm Up

For use before Lesson 14.4

1. $a = 5$ **2.** $m = 44$ **3.** $v = 12$

4. $n = 16$ **5.** $t = 6$ **6.** $k = 2.4$

14.4 Practice A

1. 49 **2.** 36 **3.** 5 **4.** 6

5. 15 **6.** 82.5 **7.** 42 oranges

8. $900 **9.** 7.5 **10.** 4 **11.** 20

12. false; *Sample answer:* If $\frac{p}{q} = \frac{3}{5}$, then $5p = 3q$ by the Cross Products Property and $\frac{5}{p} = \frac{3}{q}$ is equivalent to $3p = 5q$.

13. a. 27 cm **b.** 9 cm

14. 18 men; *Sample answer:* Because 2 out of 7 people at the lecture are men, $\frac{2}{7} = \frac{m}{63}$ where m is the number of men; $m = 18$

15. $x = 39$, $y = 25$, $z = 2.1$

16. a. 32 min **b.** 6.4 mi/h

14.4 Practice B

1. 64 **2.** 27 **3.** 350 **4.** 286

5. $42\frac{2}{3}$ **6.** 57.6 **7.** $26.64 **8.** 2

9. 17 **10.** 2.5

11. a. 5 servings **b.** 8 servings

c. 45.5 g **d.** $2\frac{1}{4}$ c

12. 121 nonmembers **13.** $h = \frac{4}{3}$ or $h = -\frac{4}{3}$

14. a. 29.19 lb **b.** 0.6 gal

Copyright © Big Ideas Learning, LLC
All rights reserved.

14.4 Enrichment and Extension

1. 1.53
2. 15 innings; not reasonable; He has been averaging about one and a half hits and walks per inning. It would be very unlikely to go 15 innings straight without allowing a hit or a walk.
3. 13 walks; Solving the proportion $\frac{1.28}{1} = \frac{92 + w}{82}$.
4. 13 hits and walks; reasonable; He has been averaging 1.28 hits and walks per inning. Because $1.28 \bullet 9 = 11.52$, he would be expected to allow between 11 and 12 hits and walks in 9 innings. So, allowing 13 hits and walks or less is reasonable.
5. 209 shots
6. The goalies faced different numbers of shots. In order to find their combined save percentage, you must find their combined number of saves and divide by their combined number of shots on goal. Solve the proportion $\frac{907}{1000} = \frac{204}{s}$ to find how many shots on goal the second goalie has faced. The second goalie has faced 225 shots. So, the goalies have made 395 total saves and had 434 total shots on goal, for a save percentage of 0.910.
7. 31 shots on goal; not reasonable; On average, he has faced about 35 shots per game and given up 3 goals per game. So it is not likely that he could make 31 saves without allowing a goal.

14.4 Puzzle Time

THE CLOCK IN THE CAFETERIA THAT WAS SLOW BECAUSE EVERY DAY AT LUNCH IT WENT BACK FOR SECONDS

14.5 Start Thinking!

For use before Activity 14.5

Sample answer: A judge compares the scores of all of the dancers. The dancer with the highest score is awarded first place.

14.5 Warm Up

For use before Activity 14.5

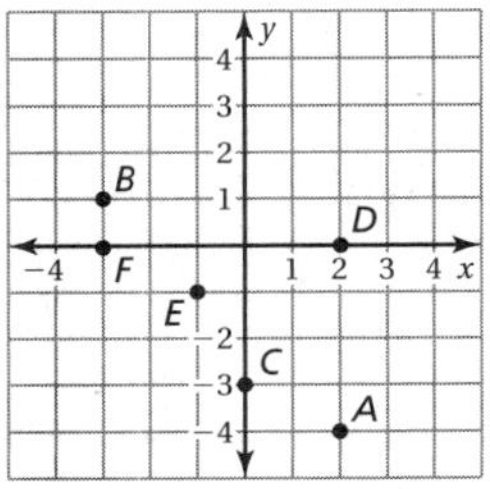

14.5 Start Thinking!

For use before Lesson 14.5

Sample answer: In auto racing, many cars travel at different speeds. They travel a distance over a period of time, which is their rate. The cars also try to maintain a consistent speed. Two or more cars can be compared on a graph to determine their speed. The car with the steepest line is traveling the fastest.

14.5 Warm Up

For use before Lesson 14.5

Time (sec)	Distance 1 (ft)	Distance 2 (ft)
0	0	0
1	12	6
2	24	12
3	36	18
4	48	24

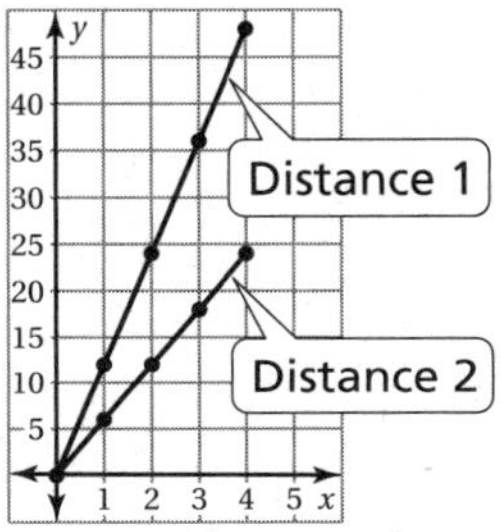

The graph for Distance 1 is steeper.

14.5 Practice A

1. 5
2. $\frac{2}{3}$
3. $\frac{3}{4}$
4. $\frac{5}{2}$
5. 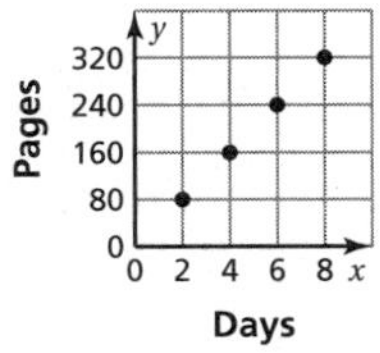

   40 pages per day
6. 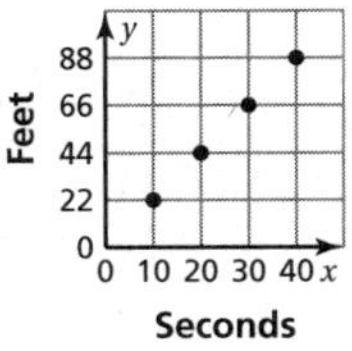

   $\frac{11}{5}$ feet per second
7. 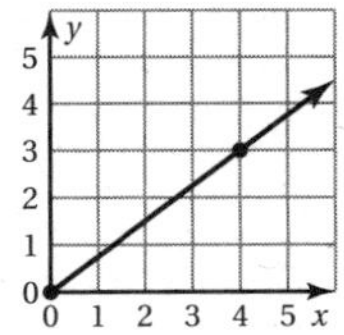

   $\frac{3}{4}$

Copyright © Big Ideas Learning, LLC
All rights reserved.

8.

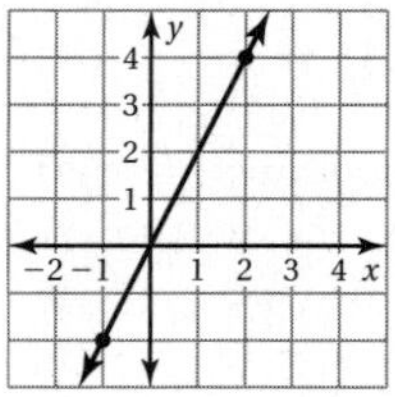

2

9.

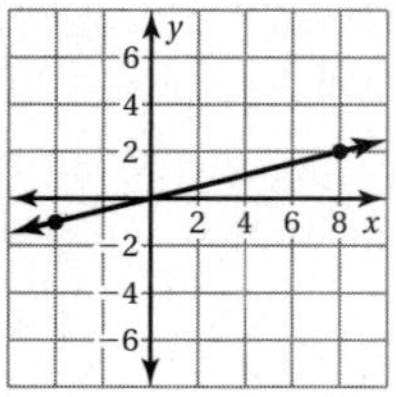

$\frac{1}{4}$

10. a. The line for the calendars is steeper than the line for the greeting cards, so a calendar is more expensive than a box of greeting cards.

b. calendar: 10; greeting cards: 6; A calendar costs \$10 and a box of greeting cards costs \$6.

c. \$6

d. 6 boxes of greeting cards or 3 calendars and 1 box of greeting cards

14.5 Practice B

1. $\frac{2}{3}$ **2.** $\frac{5}{2}$

3. $y = 3x$; For each value of x except 0, the value of $y = 3x$ is greater than $y = \frac{1}{4}x$.

4. $y = \frac{3}{5}x$; For each value of x except 0, the value of $y = \frac{3}{5}x$ is greater than $y = \frac{2}{5}x$.

5.

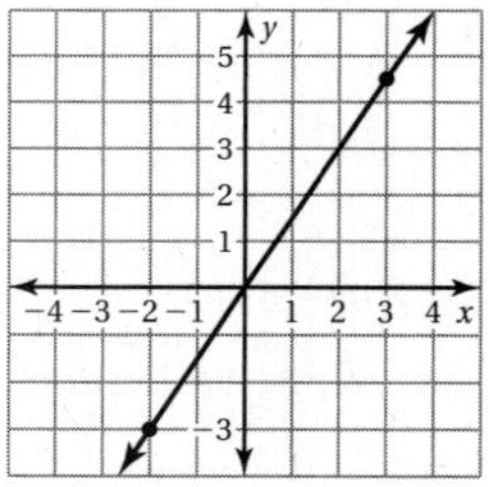

$\frac{3}{2}$

6.

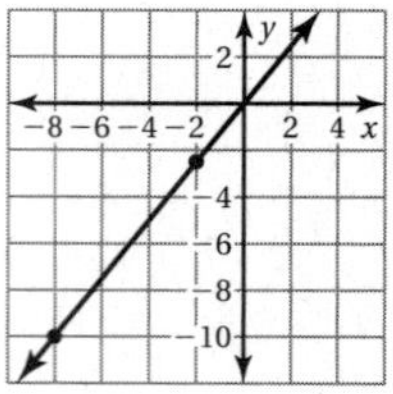

$\frac{5}{4}$

7.

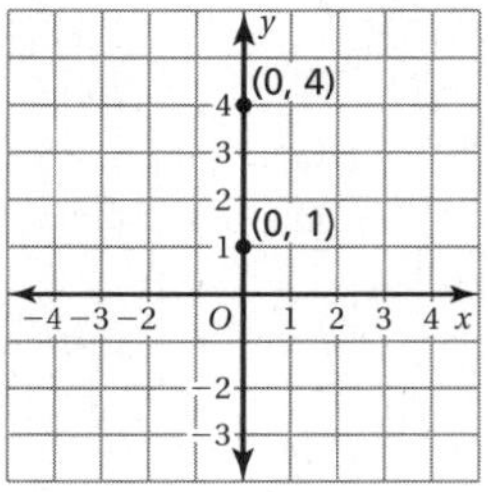

The slope is undefined because slope is the change in y divided by the change in x. The change in x is 0 in this case, and division by 0 is undefined.

8. a.

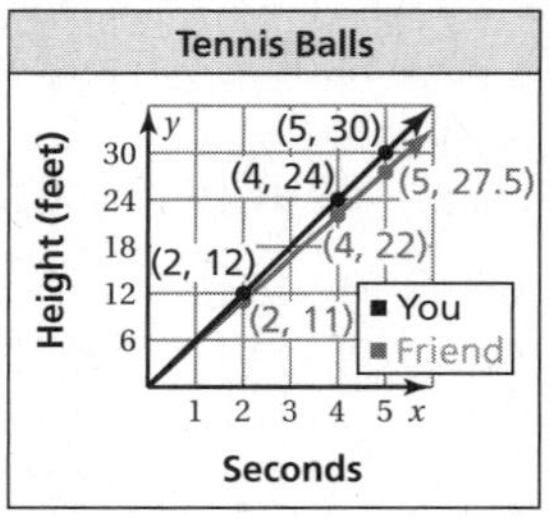

b. your ball: 6; your friend's ball: 5.5; Your ball is rising 6 ft/sec and your friend's ball is rising 5.5 ft/sec.

c. your ball; The slope is greater for the graph of the height of your tennis ball.

d. your ball: 21 ft; your friend's ball: 19.25 ft

9. 15

14.5 Enrichment and Extension

1. $\frac{1}{3}$ **2.** $\frac{5}{4}$ **3.** 1 **4.** $-\frac{3}{4}$

5. The slope would still be the same. You would obtain $\frac{-3}{-1}$, which is equal to 3.

6. *Sample answer:* $(6, 8), (8, 11), (10, 14)$

7. $y = -8$ **8.** $x = -3$

14.5 Puzzle Time

EVAPORATED MILK

14.6 Start Thinking!

For use before Activity 14.6

Sample answer: The school mascot would need to increase proportionally to your sketch. Sample dimensions for the mural could be 24 inches high and 48 inches wide.

Copyright © Big Ideas Learning, LLC
All rights reserved.

Answers for Resources by Chapter

14.6 Warm Up
For use before Activity 14.6

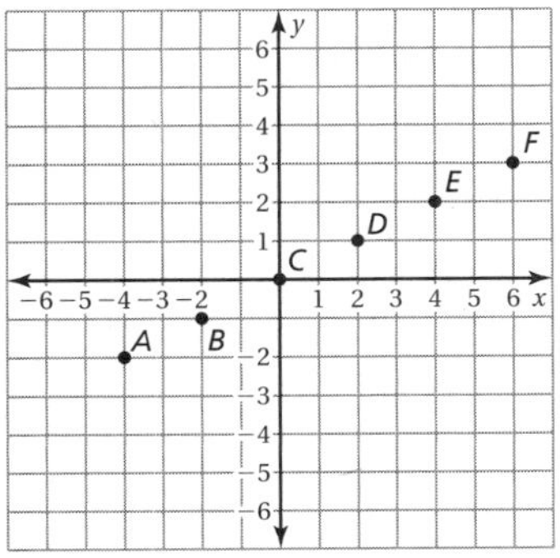

1. Start at the origin. Move 4 units left and 2 units down. Plot the point.
2. Start at the origin. Move 2 units left and 1 unit down. Plot the point.
3. Plot the point at the origin.
4. Start at the origin. Move 2 units right and 1 unit up. Plot the point.
5. Start at the origin. Move 4 units right and 2 units up. Plot the point.
6. Start at the origin. Move 6 units right and 3 units up. Plot the point.

14.6 Start Thinking!
For use before Lesson 14.6

You will need to mow your neighbor's lawn 10 times; $y = 20x$; *Sample answer:* Yes, the situation does model a direct variation because $20 stays the same (or is the constant of variation) every time you mow the lawn.

14.6 Warm Up
For use before Lesson 14.6

1. 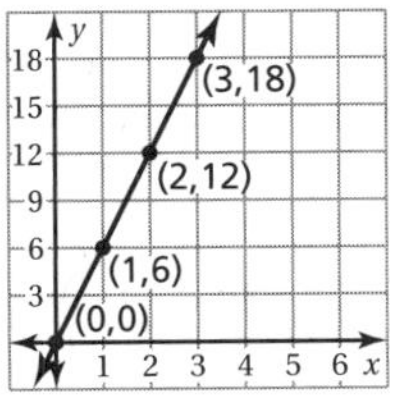

yes; All the points lie on a line and the line passes through the origin.

2. 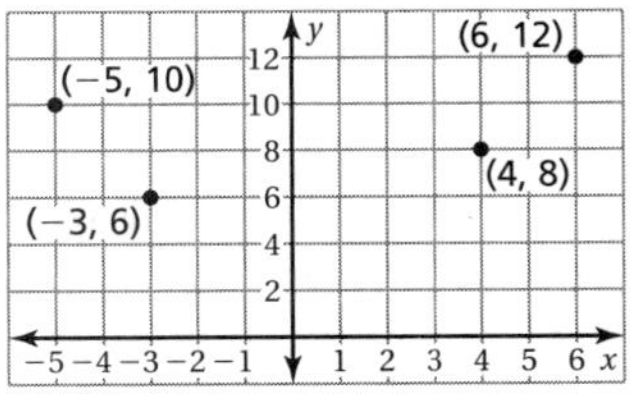

no; The points do not lie on a line.

3. 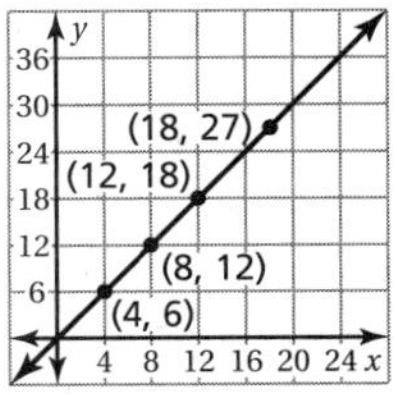

yes; All the points lie on a line and the line passes through the origin.

14.6 Practice A

1. 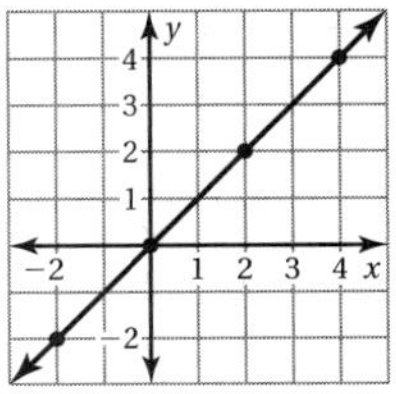

yes; The line through the plotted points passes through the origin.

2. 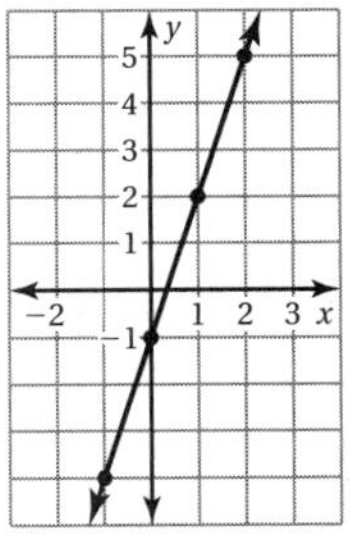

no; The line through the plotted points does not pass through the origin.

3. no; The plotted points do not lie on a line.
4. yes; The line through the plotted points passes through the origin; $k = \frac{1}{2}$
5. yes; The equation can be written as $y = kx$; $k = 3$
6. no; The equation cannot be written as $y = kx$.
7. no; The equation cannot be written as $y = kx$.

Copyright © Big Ideas Learning, LLC
All rights reserved.

8. 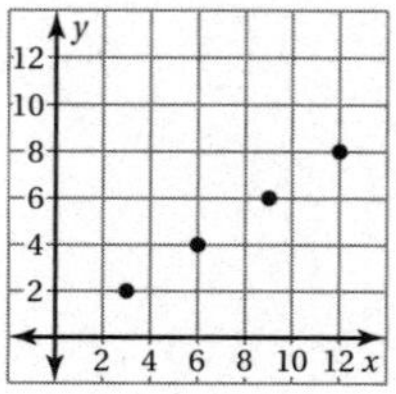

yes; $y = \frac{2}{3}x$

9. $y = 3x; k = 3$ **10.** $y = 5x; k = 5$

11. $y = 4x; k = 4$

12. a. 10 teaspoons **b.** $y = 2x$ **c.** 6 gallons

13. 7 tickets

14. a. $y = 0.95x$ **b.** $y = 3.8x$

c. $y = \frac{1}{0.95}x$, or $y = 1.05x$

d. They are reciprocals.

14.6 Practice B

1. 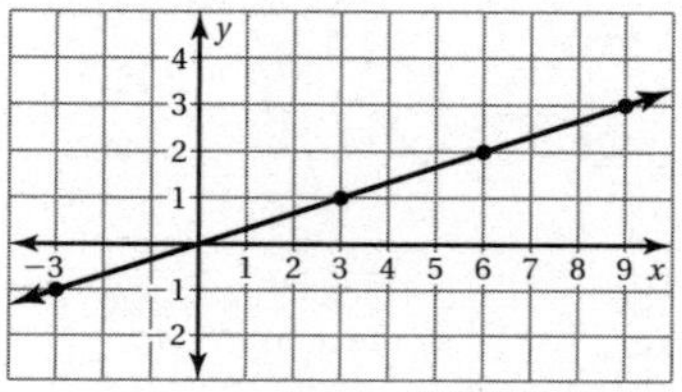

yes; The line through the plotted points passes through the origin.

2. 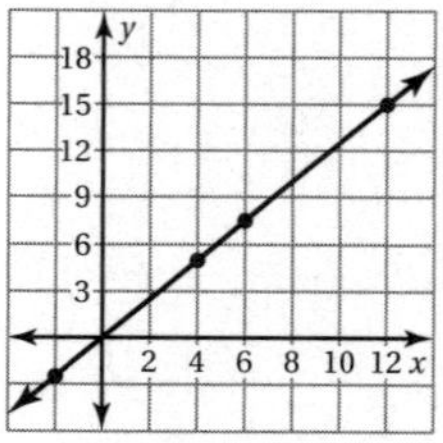

yes; The line through the plotted points passes through the origin.

3. yes; The equation can be written as $y = kx; k = 3$

4. no; The equation cannot be written as $y = kx$.

5. no; The line does not pass through the origin.

6. no; The graph is not a line.

7. a. points earned: 40, 48, 56, 64, 72, 80

b. $y = \frac{5}{4}x$

c. 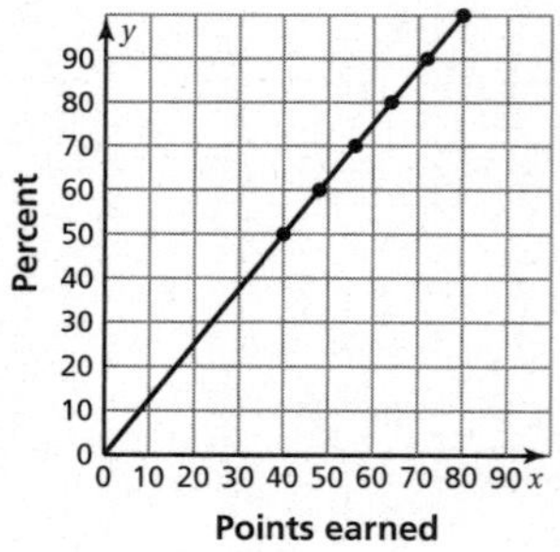

yes; The graph is a line through $(0, 0)$.

8. $y = 2x; k = 2$

9. $y = \frac{3}{2}x; k = \frac{3}{2}$ **10.** $y = \frac{11}{2}x; k = \frac{11}{2}$

11. friend's ramp

12. yes; Every direct variation has an equation of the form $y = kx$ where $k \neq 0$ and $(0, 0)$ is a solution of every equation of this form.

no; A graph that passes through the origin but is not a line does not represent a direct variation.

14.6 Enrichment and Extension

1. a. $P = -2.54n$; The store's profit is decreasing by \$2.54 for every Wacky Widget it purchases.

b. The profit is negative because the company is spending money to get the Wacky Widgets. Because k is negative, the graph will be a straight line that slopes downward from left to right.

c.

Items purchased, *n*	0	5	10
Profit, *P*	0	–\$12.70	–\$25.40

Items purchased, *n*	15	20
Profit, *P*	–\$38.10	–\$50.80

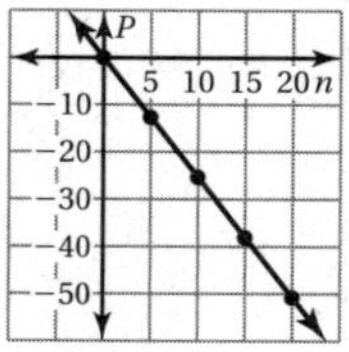

The graph makes a straight line that goes through the origin and slopes downward.

Copyright © Big Ideas Learning, LLC
All rights reserved.

Answers for Resources by Chapter

2. a. $V = 0.05p$; Tomiqua's heart pumps 0.05 liter of blood with each beat when she is at rest.

b. 4.5 L

3. a. Let m = weight on the moon, e = weight on Earth.

$$m = \frac{1}{6}e.$$

The weight on the moon is one-sixth the weight on Earth.

b. *Sample answer:* For a 150 pound person, the weight on the moon would be

$$m = \frac{1}{6}(150) = 25 \text{ lb}.$$

14.6 Puzzle Time

THEY TAKE THE BUZZ

Technology Connection

1–8. *Sample answers:* The segments could be in other positions or different lengths, but should have the same slope as those shown.

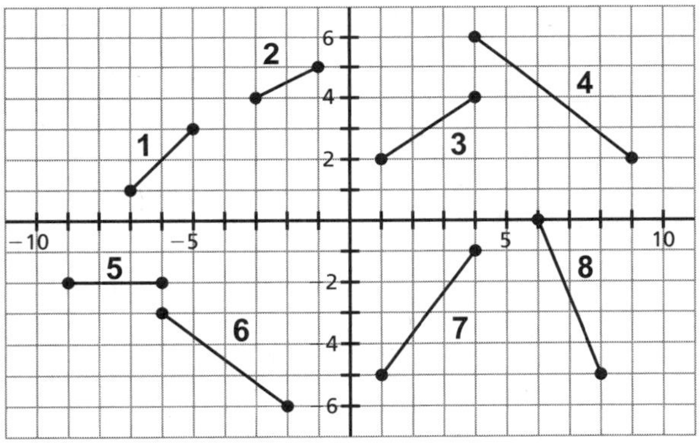

9. The slopes that are greater than 1 or less than −1 are the steepest. The slopes that are less than 1 and greater than −1 are the least steep.

10. zero

11. Vertical lines have no slope because division by zero is undefined.

12. Yes, you can move the segment around the graph without changing the slope by not moving the endpoints. The segments are parallel to one another.

Chapter 15

15.1 Start Thinking!

For use before Activity 15.1

Sample answer: The percent has the decimal point two places to the right; no; Andrew only moved the decimal point one place to the right. It should be 98%.

15.1 Warm Up

For use before Activity 15.1

1. $\frac{3}{8}$ **2.** $\frac{1}{2}$ **3.** $\frac{1}{4}$

4. $\frac{1}{8}$ **5.** $\frac{5}{6}$ **6.** $\frac{3}{8}$

15.1 Start Thinking!

For use before Lesson 15.1

Sample answer: In basketball, free throw percentage, field goal percentage, and three point percentage are often given as decimals, which makes them easier to compare.

15.1 Warm Up

For use before Lesson 15.1

1. 0.19 **2.** 0.02 **3.** 0.075

4. 89% **5.** 54% **6.** 10%

15.1 Practice A

1. 0.81 **2.** 0.78 **3.** 0.05 **4.** 0.08

5. 0.4 **6.** 0.6 **7.** 0.237 **8.** 0.1675

9. 1.5 **10.** 2.1 **11.** 1.86 **12.** 4.16

13. 1.008 **14.** 0.0517 **15.** 0.004 **16.** 0.0004

17. The decimal point is moved to the right instead of the left.

1.475% = 01.475% = 0.01475

18. 66% **19.** 32% **20.** 51% **21.** 97%

22. 1% **23.** 4% **24.** 31.2% **25.** 46.8%

26. 50% **27.** 120% **28.** 108% **29.** 116%

30. 0.3% **31.** 2.5% **32.** 2.45% **33.** 202.5%

34. The decimal point is moved from the wrong place.

1.8 = 1.80 = 180%

35. 0.54 **36.** 15% **37.** 88%

Copyright © Big Ideas Learning, LLC
All rights reserved.

38. 0.86, 86% **39.** 5 times **40.** $\frac{21}{100}$; 0.21

41. $\frac{3}{4}$, 0.75 **42.** $\frac{16}{25}$, 0.64 **43.** $\frac{17}{20}$, 0.85

15.1 Practice B

1. 54% **2.** 37% **3.** 22.2% **4.** 92.9%

5. 140% **6.** 250% **7.** 2000% **8.** 0.5%

9. $\frac{17}{25}$, 0.68 **10.** $\frac{9}{100}$, 0.09 **11.** $\frac{11}{20}$, 0.55

12. $\frac{13}{50}$, 0.26 **13.** $\frac{53}{125}$, 0.424 **14.** $\frac{92}{125}$, 0.736

15. $\frac{5}{16}$, 0.3125 **16.** $\frac{893}{2000}$, 0.4465

17. 64%

18. a. 0.48, $\frac{12}{25}$; 0.28, $\frac{7}{25}$; 0.14, $\frac{7}{50}$ **b.** 10%

c.

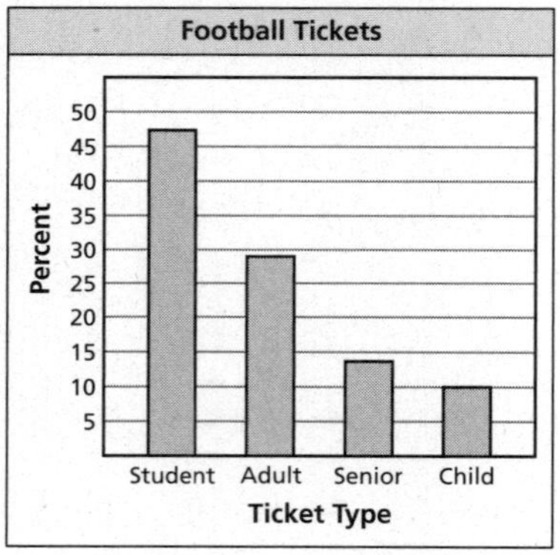

19. a. 33% **b.** 35% **c.** $1\frac{3}{4}$ times **d.** 27%, 0.27

20. a–d.

Distance	within 1 mi	1–2 mi	2–3 mi	over 3 mi
Percent	50%	28%	10%	12%

15.1 Enrichment and Extension

1. 50%; 0.5 **2.** 50%; 0.5

3. a. 25%; 0.25 **b.** 75%; 0.75
c. 50%; 0.5 **d.** 50%; 0.5

4. a. 56.25%; 0.5625 **b.** 43.75%; 0.4375

5. a. 4 **b.** 9; 28.125%, 0.28125

15.1 Puzzle Time

BOTH ARE SUPPOSED TO BE DUE

15.2 Start Thinking!

For use before Activity 15.2

Sample answer: You can rewrite each fraction as a decimal and compare decimals or you can write each fraction using the same denominator and compare the numerators. $\frac{7}{12}$; methods may vary.

15.2 Warm Up

For use before Activity 15.2

1. 1.4, 1.44, 3.3, 4.1, 4.3

2. 0.5, 0.7, 1.5, 1.57, 1.75

3. 0.02, 0.04, 0.2, 0.24, 0.44

4. 0.06, 0.6, 6.06, 6.6, 6.63

5. 0.02, 0.17, 0.2, 0.72, 2.27

6. 0.18, 0.81, 1.8, 1.88, 8.1

15.2 Start Thinking!

For use before Lesson 15.2

Answers will vary. Check students' work.

15.2 Warm Up

For use before Lesson 15.2

1. 45% **2.** 0.4 **3.** $\frac{9}{17}$

4. 0.6 **5.** $\frac{2}{5}$ **6.** 240%

15.2 Practice A

1. $\frac{3}{4}$ **2.** 0.54

3. neither; 0.21 = 21% **4.** $\frac{2}{3}$

5. 49% **6.** 16%

7. neither; $\frac{12}{25} = 48\%$ **8.** 12%

9. 1.2 **10.** $\frac{31}{50}$ **11.** $50\frac{1}{4}$ **12.** $\frac{1}{8}$

13. $\frac{68}{100} = 68\%$, not 0.68%.

So, $68\% > 0.7\%$ and $\frac{17}{25}$ is the greater number.

Copyright © Big Ideas Learning, LLC
All rights reserved.

14. 63%, 0.64, $\frac{13}{20}$

15. $\frac{11}{25}$, 45%, 0.46

16. 0.12, $\frac{1}{8}$, 13%, 0.135

17. $\frac{7}{8}$, 90%, 0.925, 0.93, $\frac{15}{16}$

18. $3\frac{3}{5}$, 362%, 3.66, $3\frac{2}{3}$, 36

19. 0.27, 27.3%, $\frac{11}{40}$, 28%, 0.3

20. the punch that you made

21. You made a higher percent of baskets.

22. yes; 62.5% is greater than 60%.

23. grade 7, grade 6, grade 8

15.2 Practice B

1. $\frac{1}{4}$ **2.** $\frac{5}{9}$ **3.** 3.2 **4.** 1

5. 0.3, $\frac{8}{25}$, 33%, $\frac{1}{3}$, 33.6%

6. 210%, $\frac{43}{20}$, 2.2, $2.\overline{2}$

7. B **8.** C **9.** A **10.** D

11. *Sample answer:* Write 31% and 32% as the decimals 0.31 and 0.32. Graph 0.31 and 0.32 on a number line. Then identify a decimal between them, such as 0.315.

12. yes; 5% of a meter $= 0.05(100 \text{ cm}) = 5 \text{ cm}$;
Because 6 centimeters is greater than 5 centimeters, 6 centimeters is greater than 5% of a meter.

13. no; 6% of 1 lb $= 0.06(16 \text{ oz}) = 0.96 \text{ oz}$;
Because 0.96 ounce is less than 1 ounce, 6% of a pound is less than an ounce.

14. 0.0004 of a day, 1% of an hour, $\frac{2}{3}$ of a minute

15. a. 0.45%, 0.0082, $\frac{1}{36}$, $\frac{1}{16}$

b. The population of Florida has grown faster than the population of the U.S.; *Sample answer:* People tend to move to Florida because it is a nice place to live.

c. *Sample answer:* no; Some day Florida will become too crowded to keep growing.

16. no;

$$0.000002\% = 0.00000002 = \frac{2}{100{,}000{,}000} = \frac{20}{1{,}000{,}000{,}000} = 20 \text{ parts per billion.}$$

Because this is greater than 10 parts per billion, it is not an allowable amount.

15.2 Enrichment and Extension

1%, $\frac{11}{40}$, $\frac{17}{60}$, 31%, $\frac{9}{25}$, $0.3\overline{81}$, $\frac{31}{75}$, 41.5%, $\frac{56}{99}$, $0.\overline{75}$, $\frac{101}{120}$, 85.25%, 100%

15.2 Puzzle Time

IT WAS TRULY IN A JAM

15.3 Start Thinking!

For use before Activity 15.3

Sample answer: Percents are used in banks to calculate interest rates.

15.3 Warm Up

For use before Activity 15.3

1. *Sample answer:* 260 **2.** *Sample answer:* 220

3. *Sample answer:* −140 **4.** *Sample answer:* −1200

5. *Sample answer:* −19 **6.** *Sample answer:* 500

15.3 Start Thinking!

For use before Lesson 15.3

First determine how many goals your team scored, which is 10 goals. Then draw a model. Your team scored 60% of its goals during regular play.

15.3 Warm Up

For use before Lesson 15.3

1. 12 **2.** 50% **3.** 50

4. 104 **5.** 72% **6.** 80

Copyright © Big Ideas Learning, LLC
All rights reserved.

Answers for Resources by Chapter

15.3 Practice A

1. 8 2. 40% 3. 70 4. 138

5. $\frac{15}{w} = \frac{40}{100}$; 37.5 6. $\frac{24}{w} = \frac{0.6}{100}$; 4000

7. $\frac{a}{100} = \frac{27}{75}$; 36% 8. $\frac{17}{68} = \frac{p}{100}$; 25%

9. 48 10. $450

11. $\frac{0.2}{16} = \frac{p}{100}$; 1.25% 12. $\frac{19.6}{w} = \frac{24.5}{100}$; 80

13. $\frac{\frac{3}{5}}{w} = \frac{30}{100}$; 2 14. $\frac{a}{\frac{5}{9}} = \frac{45}{100}$; $\frac{1}{4}$

15. 7 left to complete

16. a. 60% b. 36% c. 48%

15.3 Practice B

1. $\frac{33}{w} = \frac{55}{100}$; 60 2. $\frac{42}{120} = \frac{p}{100}$; 35%

3. $\frac{36}{w} = \frac{0.8}{100}$; 4500 4. $\frac{48}{64} = \frac{p}{100}$; 75%

5. 72 6. $\frac{3.69}{w} = \frac{90}{100}$; $4.10

7. $\frac{6}{w} = \frac{6.25}{100}$; 96 8. $\frac{\frac{7}{8}}{w} = \frac{70}{100}$; 1.25

9. $\frac{7.2}{w} = \frac{250}{100}$; 2.88 10. $\frac{a}{\frac{3}{8}} = \frac{72}{100}$; 0.27

11. $\frac{1.4}{1.12} = \frac{p}{100}$; 1.25% 12. $\frac{86.8}{w} = \frac{140}{100}$; 62

13. $\frac{1}{2}x$

14. a. $31.5x$ b. 56.25%

15.3 Enrichment and Extension

1. $x = 4$ 2. $x = 8$ 3. $x = 4$ 4. $x = 17$

15.3 Puzzle Time

HOLDUP IN THE YARD WHEN TWO CLOTHESPINS HELD UP A PAIR OF PANTS

15.4 Start Thinking!

For use before Activity 15.4

36 points; *Sample answer:* The percent proportion is $\frac{a}{w} = \frac{p}{100}$. Substitute 45 for w and 80 for p. Solve for a. You need to get 36 points.

15.4 Warm Up

For use before Activity 15.4

1. 40% 2. 55% 3. 33

4. 18.9 5. 85 6. 1

15.4 Start Thinking!

For use before Lesson 15.4

The percent equation is correct. The number is 1.5. *Sample answer:* The student put the 30 as the part instead of the whole in the percent proportion.

15.4 Warm Up

For use before Lesson 15.4

1. 18 2. $33\frac{1}{3}\%$ 3. 40

4. 96 5. 45% 6. 140

15.4 Practice A

1. 40%; Methods will vary.

2. 20; Methods will vary.

3. $a = \frac{70}{100} \bullet 120$; 84 4. $30 = p \bullet 120$; 25%

5. $112 = \frac{56}{100} \bullet w$; 200 6. $128 = p \bullet 80$; 160%

7. $a = \frac{140}{100} \bullet 45$; 63 8. $15 = \frac{6}{100} \bullet w$; 250

9. 21 competitors 10. 30 students

11. 75%

12. a. 38 pennies b. 95 rare coins c. 57 nickels

13. a. 6% b. $7.50 c. $12.50

14. true; To find 120% of a number, multiply it by $\frac{120}{100}$, which is a number greater than 1.

15. false; To find 0.5% of a number, multiply it by $\frac{0.5}{100}$, or $\frac{5}{1000}$.

Copyright © Big Ideas Learning, LLC
All rights reserved.

15.4 Practice B

1. 30%; Methods will vary.
2. 140; Methods will vary.
3. $27 = \frac{0.5}{100} \bullet w$; 5400
4. $a = \frac{125}{100} \bullet 240$; 300
5. $28 = \frac{1.4}{100} \bullet w$; 2000
6. $27 = p \bullet 72$; 37.5%
7. 9.44 in.
8. 20,000 gal
9. yes
10. a. 320 students b. 80 students
    c. 32 students d. 64 students
11. 10.8
12. 1000 and 1500
13. true; $A = \frac{45}{100}B$, so $\frac{A}{B} = \frac{45}{100} = \frac{9}{20}$.
14. 42.11%
15. 131.58%

15.4 Enrichment and Extension

1. yes; You should draw again, because you are currently over 50%.
2. The lowest card you could draw is 13. The highest card you could draw is 19. You could still win with 35% of the cards.
3. The most you can get on your first three draws and still have a chance to win is 39. If you have this score after your first three draws, your only chance of winning is to get a 1. Otherwise you will be over 50%.
4. *Sample answer:* It is probably best not to draw again. You have 35%. So, only 6 out of 20 cards will increase your percent without causing you to go over. If you draw an 8 or 9 first, it would be a good idea to stop. Your current score is even harder to beat, and you are even less likely to draw a card that will increase your percent without making you go over.
5. *Sample answer:* If the target percent is higher, players will be less likely to go over. So, you will have to get closer to the target percent in order to win. If the target percent is lower, players will be more likely to go over. So, you may not have to get as close to the target percent in order to win.

15.4 Puzzle Time

THE SPORTS FAN WHO LISTENED TO A MATCH AND ENDED UP BURNING HIS EAR

15.5 Start Thinking!

For use before Activity 15.5

Sample answer: A meteorologist uses percents to tell you the chance of precipitation for a given day or the amount of humidity in the air.

15.5 Warm Up

For use before Activity 15.5

1. 45%
2. 134%
3. 54.9%
4. 108%
5. 98.5%
6. 32.25%

15.5 Start Thinking!

For use before Lesson 15.5

Sample answer: You have $250 worth of baseball cards. You are told that the value of the baseball cards will decrease 35% over the next 6 months. How much will the baseball cards be worth at the end of the 6 months?

15.5 Warm Up

For use before Lesson 15.5

1. 18 inches
2. 58 gallons
3. 126 meters
4. 39 grams
5. 114 pounds
6. 50 liters

15.5 Practice A

1. 9 dogs
2. 203 fluid ounces
3. 199 textbooks
4. 15 students
5. increase; 60%
6. decrease; 10%
7. decrease; 28.6%
8. decrease; 17.9%
9. increase; 100%
10. decrease; 50%
11. 20% increase
12. a. decrease of 20% b. 173 tickets
    c. 35.9% decrease
13. 4 musicians
14. a. 10 feet b. 25% increase c. 56.3% increase

15.5 Practice B

1. 66 employees
2. 10°
3. 45 customers
4. 4221 fans
5. decrease; 25%
6. increase; 107.3%
7. increase; 100%
8. decrease; 60%
9. 3.7% decrease
10. a. 2% decrease b. 1152 hamburgers
11. a. $36.00 b. increase of 4.2%

Copyright © Big Ideas Learning, LLC
All rights reserved.

12. **a.** 14.3% increase **b.** 14.3% increase
    c. Troop B
    d. neither troop; The percent of change is the same.

15.5 Enrichment and Extension

1. **a.** 60 thousand pairs **b.** 75 thousand pairs
2. **a.** 172 thousand pairs **b.** 86 thousand pairs
3. You should use percent of increase because your company's percent of increase is bigger than your competitor's percent of increase, even though your competitor increased their sales by more pairs of sneakers.
4. **a.** 297.5 thousand pairs **b.** 122.5 thousand pairs
5. **a.** 270 thousand pairs **b.** 120 thousand pairs
6. You should use amount of increase because your company increased their sales by more pairs than your competitor, even though their percent of increase was greater.

15.5 Puzzle Time

SOMEBODY FRAMED IT

15.6 Start Thinking!

For use before Activity 15.6

Sample answer: A store marks up an item so they can make a profit. A store may then discount the item because either it did not sell or they need the space for new items.

15.6 Warm Up

For use before Activity 15.6

1. $a = 0.4 \bullet 238$; 95.2
2. $28 = p \bullet 70$; 40%
3. $a = 0.34 \bullet 240$; 81.6
4. $6 = 0.05 \bullet w$; 120
5. $a = 1.1 \bullet 150$; 165
6. $42 = 2.5 \bullet w$; 16.8

15.6 Start Thinking!

For use before Lesson 15.6

The jeans that cost $35 with a 30% discount are a better bargain. They cost $24.50 on sale. The other pair of jeans cost $26 on sale.

15.6 Warm Up

For use before Lesson 15.6

1. $23
2. $84.50
3. $112
4. $157.50

15.6 Practice A

1. $52.50
2. $6.30
3. $60
4. $200
5. 20%
6. 15%
7. $81.25
8. $68
9. 40%
10. **a.** $192 **b.** $134.40 **c.** make money; $14.40
11. $84
12. **a.** $63.60 **b.** $51
    c. $54.06 **d.** decrease of 15%

15.6 Practice B

1. $71.50
2. 65%
3. $72
4. $2128
5. 40%
6. $350
7. **a.** $31.94 **b.** $36.73 **c.** yes; $33.05
8. **a.** $37.99 **b.** 39% **c.** $27.99 **d.** 54%
9. 67%

15.6 Enrichment and Extension

1. **a.** $15.40 **b.** $5.50
   c. The coupon takes 30% off of the sale price, not the original price. So, you are saving less than 90% off the original.
2. **a.** In order to find the amount of markup, you find 40% of the cost. In order to find the amount of discount, you find 40% of the selling price. Because the selling price is larger than the cost, the amount of discount would be larger than the amount of markup. So, you would be discounting them by more than you marked them up, and you would lose money.
   b. $125.36 **c.** 29%
3. **a.** Yes, you would be allowed to match the competitor's price. You would be discounting the player by $9.20, which is just under a 3% discount.
   b. No, you would not be allowed to match the Internet price. The one on the Internet would cost $62.97 with shipping and handling. So, you would have to discount the router by $11.78, which is a percent of discount of 15.8%. The least amount that you can offer on the router is $63.54.

15.6 Puzzle Time

PLEASED TO EAT YOU

15.7 Start Thinking!

For use before Activity 15.7

Sample answer: Pro: Build credit; Con: Could lead to debt you cannot repay

15.7 Warm Up

For use before Activity 15.7

1. $225
2. $45
3. $1040
4. $80.75
5. $652.50
6. $549.25

Copyright © Big Ideas Learning, LLC
All rights reserved.

Answers for Resources by Chapter

15.7 Start Thinking!

For use before Lesson 15.7

Sample answer: Some factors you must consider are interest rate or if the bank charges a fee for a minimum balance.

15.7 Warm Up

For use before Lesson 15.7

1. a. $45 **b.** $795 **2. a.** $36 **b.** $336

3. a. $280 **b.** $1680 **4. a.** $189 **b.** $789

5. a. $22 **b.** $572 **6. a.** $21 **b.** $1221

15.7 Practice A

1. a. $30 **b.** $230 **2. a.** $120 **b.** $870

3. a. $80 **b.** $1680 **4. a.** $30 **b.** $530

5. 2% **6.** 7% **7.** 3 years **8.** 1.5 years

9. $367.50 **10.** $1400 **11.** $4200 **12.** 18 years

13. a. $300 **b.** $3300

14. a. $3360 **b.** $140 **c.** credit card

15. a. $1296 **b.** $1399.68

15.7 Practice B

1. a. $332.80 **b.** $2932.80

2. a. $1593.75 **b.** $76,593.75

3. 4.75% **4.** 6.5% **5.** 6 months **6.** 3 years

7. $35,000 **8.** $7800 **9.** 9% **10.** 10%

11. a. $1208.10 **b.** $1220.18 **c.** $1228.11
d. $28.11 **e.** 2.3%

12. 8%

15.7 Enrichment and Extension

1. If she gets $500 off, the total cost would be $21,320.57. If she decreases the interest rate, the total cost would be $21,519.48. It would be better to get $500 off.
2. The total cost would be $22,278.63. When interest is compounded more often, the total cost increases, which means that the lender is making more money.
3. *Answer should include, but is not limited to:* The student must include the car's original price, the interest rate, how often the interest is compounded, the total cost of the loan, and the monthly payment.

15.7 Puzzle Time

ARE YOU ASLEEP

Technology Connection

1. 273 **2.** 10 **3.** 27

4. 67.5 **5.** 300 **6.** 700

7. 15.5 **8.** 400 **9.** 5

10. $5,200

Copyright © Big Ideas Learning, LLC
All rights reserved.

Answers for Assessment Book

Chapter 11

11.1–11.3 Quiz

1. 8 **2.** 3 **3.** $<$

4. $<$ **5.** $>$ **6.** $=$

7. $-4, -2, 1, 3, |-5|$ **8.** $-5, 0, |8|, |-9|, 11$

9. -7 **10.** 2 **11.** -10

12. -7 **13.** 11 **14.** 4

15. -9 feet **16.** 61 points **17.** -1

11.4–11.5 Quiz

1. negative **2.** negative **3.** positive **4.** positive

5. -40 **6.** 12 **7.** -63 **8.** -3

9. -4 **10.** -9 **11.** -8 **12.** 15 people

13. -24 **14. a.** \$7 **b.** \$56

15. a. -14 **b.** -2 gallons **c.** 35 miles per gallon

Test A

1. 1 **2.** 15 **3.** 37 **4.** $=$

5. $<$ **6.** $<$ **7.** $-4, -3; -3$

8. 3 **9.** 4 **10.** 13 **11.** 0

12. -28 **13.** -30 **14.** 3 **15.** -3

16. -2 **17.** 0

18. 38 meters below base camp

19. 7°C **20.** -3 **21.** -40 **22.** 0

23. $m = 14$ **24.** $w = 4$ **25.** $c = 16$

26. $k = -6$ **27.** $n = 13$ **28.** $d = -6$

29. $96, -192$ **30.** $-20{,}000, 200{,}000$

31. a. 22°F; -15°F **b.** 37°F **c.** -21°F **d.** 0°F

e. *Sample answer:* You do not know what the temperature was at other points during the day.

Test B

1. 12 **2.** 30 **3.** 44

4. $>$ **5.** $=$ **6.** $<$

7. $-4, |0|, |-3|, |-5|, 7$

8. $-6, -2, |-6|, |-7|, 10$

9. Duluth **10.** -186 **11.** -21 **12.** 53

13. -37 **14.** 180 **15.** -210 **16.** 13

17. undefined **18.** -2 feet

19. 23; The pairs of integer factors of -24 are 1 and -24, -1 and 24, 2 and -12, -2 and 12, 3 and -8, -3 and 8, 4 and -6, and -4 and 6. Of these, -1 and 24 have the greatest sum.

20. -8 **21.** -45 **22.** -33 **23.** -7

24. a. lowest, -13; highest, 73 **b.** 31

25. -1; The sum of the numbers must be 20. Because $21 + (-1)$ is 20, the number is -1.

26. $m = -23$ **27.** $d = -26$

28. $324, -972$ **29.** $-50, 25$

30. a. Friday **b.** 180 **c.** -34 **d.** 27

31. *Sample answer:* $a = 2, b = -1$
(The statement is false if a or b are negative.)

Alternative Assessment

1. a. *Sample answer:* To help Alec, I will first use the following problems.

$|7|$ and $|-4|$

We will use a number line for the absolute value problems.

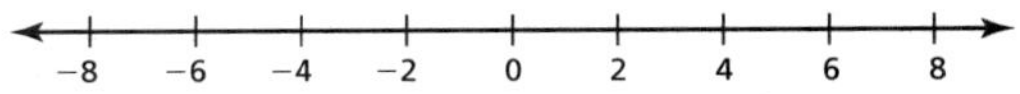

For $|7|$, Alec will graph 7. I will explain to him that the distance between 7 and zero is 7. So, $|7| = 7$.

For $|-4|$, Alec will graph -4. I will explain to him that the distance between -4 and zero is 4. So, $|-4| = 4$.

b. *Sample answer:* To help Alec with operations with integers, I will use the following problems.

$6 + (-2)$ $6 - (-2)$ $6 \times (-2)$ $-6 \div 2$

Copyright © Big Ideas Learning, LLC
All rights reserved.

c. *Sample answer:* We will use positive counters and negative counters for all of the problems. For $6 + (-2)$, Alec will start with 6 positive counters. I will explain that because he is adding -2, he will add 2 negative counters to his 6 positive counters. Then I will explain that each pair of positive and negative counters is equal to zero. We will remove two pairs of positive and negative counters. Because Alec is left with 4 positive counters, $6 + (-2) = 4$.

For $6 - (-2)$, Alec will start with 6 positive counters then he will take away 2 negative counters. Because Alec does not have any negative counters, I will explain that he needs to add two pairs of positive and negative counters. This does not change the value of the problem because each pair is equal to zero. So instead of 6 positive counters, we have 8 positive counters and 2 negative counters, which is still equal to 6. Now we can find $6 - (-2)$ by removing 2 negative counters. Because Alec is left with 8 positive counters, $6 - (-2) = 8$.

For $6 \times (-2)$, Alec will make 6 piles, each with 2 negative counters. Because Alec has 12 negative counters, $6 \times (-2) = -12$.

For $-6 \div 2$, Alec will start with 6 negative counters. I will explain that he needs to divide the counters into 2 piles. Because there are 3 negative counters in each pile, $-6 \div 2 = -3$.

d. *Sample answer:* We will use the following number line for the addition, subtraction, and division problems. I will encourage Alec to visualize himself standing on the number line.

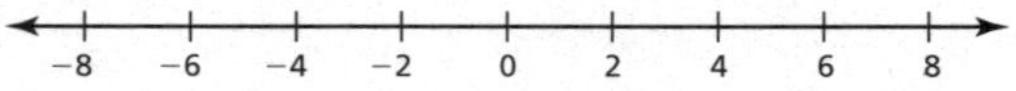

For $6 + (-2)$, Alec will start at 6. I will explain that because we are adding, he should "face" in the positive direction. Because we are adding a negative number, he will take two steps backward and stop at 4. So, $6 + (-2) = 4$.

For $6 - (-2)$, Alec will start at 6. I will explain that because we are subtracting, he should "face" in the negative direction. Because we are subtracting a negative number, he will take two steps backward and stop at 8. So, $6 - (-2) = 8$.

For the multiplication, I will ask Alec to draw a number line from -16 to 16. It should look like this:

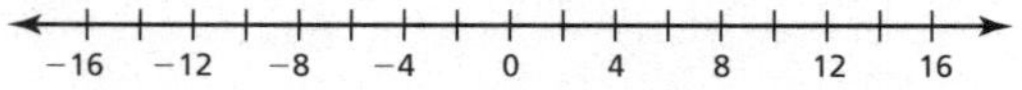

For $6 \times (-2)$, Alec will start at 0. I will explain that because the first number is positive, he should "face" in the positive direction. Because the second number is negative, he will make 6 jumps of 2 backwards and stop at -12.

So, $6 \times (-2) = -12$.

For $-6 \div 2$, Alec will start at -6 and "face" in the negative direction. I will explain that he needs to divide the distance between -6 and 0 into 2 equal line segments. Because Alec is still facing in the negative direction, the answer is negative. So, $-6 \div 2 = -3$.

Copyright © Big Ideas Learning, LLC
All rights reserved.

Answers for Assessment Book

e. *Sample answer:*

Name Rule or Definition	Sample Problems
Absolute Value The distance from zero on a number line.	$\lvert 7 \rvert = 7$ $\lvert -4 \rvert = 4$
Addition (both positive) Add. The sum is positive.	$4 + 3 = 7$
Addition (both negative) Add. The sum is negative.	$-4 + (-3) = -7$
Addition (positive and negative) Subtract the number with the lesser absolute value from the number with the greater absolute value and use the sign of the number with the greater absolute value.	$6 + (-2) = 4$
Subtraction Follow the addition rules to add the opposite of the number being subtracted.	$6 - (-2) = 6 + 2$ $= 8$
Multiplication (same signs) Multiply. The product is positive.	$2(5) = 10$ $-2(-5) = 10$
Multiplication (different sign) Multiply. The product is negative.	$6 \times (-2) = -12$
Division (same signs) Divide. The quotient is positive.	$10 \div 2 = 5$ $-10 \div (-2) = 5$
Division (different signs) Divide. The quotient is negative.	$-6 \div 2 = -3$

Chapter 12

12.1–12.2 Quiz

1. -0.9 **2.** -1.6 **3.** $-\frac{13}{20}$ **4.** $-3\frac{3}{4}$

5. = **6.** < **7.** > **8.** >

9. $-2\frac{1}{8}$ **10.** $-4\frac{1}{12}$ **11.** -10.9

12. 3.535 **13.** Town B **14.** \$84.92

12.3–12.4 Quiz

1. -3.2 **2.** 6.8 **3.** $-7\frac{5}{6}$ **4.** $-1\frac{7}{18}$

5. 8 **6.** $4\frac{2}{5}$ **7.** -27.9 **8.** $-1\frac{1}{3}$

9. 15 **10.** -3 **11.** 3.1 **12.** $-3\frac{8}{15}$

13. 1.69 fl oz **14.** $\frac{19}{40}$ **15.** $-\$13.30$

Test A

1. $-0.\overline{27}$ **2.** -3.65 **3.** $3\frac{21}{50}$ **4.** $-\frac{7}{20}$

5. < **6.** > **7.** your friend's

8. 9.29 **9.** $\frac{2}{3}$ **10.** $10\frac{9}{10}$ **11.** -12.13

12. 4.3 **13.** $3\frac{1}{2}$ **14.** $\frac{2}{3}$ **15.** \$7.84

16. $\frac{1}{2}$ **17.** -0.09 **18.** -0.35

19. $\frac{3}{4}$ **20.** $2\frac{1}{4}$ **21.** -3.84

22. 12 packages; $8 \div \frac{2}{3} = 8 \bullet \frac{3}{2} = 12$

23. 31 in. **24.** 2.7 gallons

Test B

1. $-1.6, -\frac{4}{3}, \frac{2}{3}, \frac{4}{5}, 0.9$

2. > **3.** = **4.** 15.9 **5.** $7\frac{7}{24}$

6. 3.265 **7.** -2.63 **8.** $-6\frac{1}{9}$ **9.** $8\frac{3}{20}$

10. $-1\frac{1}{24}$ **11.** $-2\frac{7}{24}$ **12.** $-2\frac{11}{12}$ **13.** $3\frac{13}{24}$

14. a. $4\frac{5}{12}$ hours **b.** $2\frac{11}{12}$ hours

15. *Sample answer:* $-2.\overline{27}$

Copyright © Big Ideas Learning, LLC
All rights reserved.

Answers for Assessment Book

16. $5\frac{7}{10}$ **17.** 7.09 **18.** $-12\frac{1}{4}$ **19.** -3.45

20. $1\frac{1}{12}$ **21.** -5.81 **22.** -1.08 inches

23. no; $\frac{1}{6}$ cup; $2\frac{1}{2} = 2\frac{3}{6}$ and $2\frac{1}{3} = 2\frac{2}{6}$, so $2\frac{1}{2} > 2\frac{1}{3}$; $2\frac{1}{2} - 2\frac{1}{3} = \frac{1}{6}$, so you need $\frac{1}{6}$ cup more sugar.

24. 3.5 gallons **25.** 7 packages

Alternative Assessment

1. a. *Sample answer:* $-\frac{7}{8}, -\frac{4}{9}, \frac{3}{4}$

b. *Sample answer:*

$\frac{3}{4} + \left(-\frac{7}{8}\right), -\frac{4}{9} - \frac{3}{4}, -\frac{7}{8} - \left(-\frac{4}{9}\right), -\frac{4}{9} + \frac{3}{4};$

$-\frac{4}{9} \times \frac{3}{4}, -\frac{7}{8}\left(-\frac{4}{9}\right), -\frac{4}{9} \div \frac{3}{4}, \frac{3}{4} \div \left(-\frac{7}{8}\right);$

$-\frac{4}{9} \times \frac{3}{4}$: The water height of a tank decreases by $\frac{4}{9}$ yard every hour. What is the change in height after 45 minutes?

c. *Sample answer*: $-6.5, 0.25, 0.75;$

$0.75 + 0.25, 0.75 - (-6.5), -6.5 + 0.25, 0.25 - 0.75;$

$0.75 \div 0.25, 0.75(-6.5), -6.5 \div 0.75, 0.25 \times (-6.5);$

$0.75 \div 0.25$: I have 0.75 pound of tuna salad, and I want to know how many sandwiches I can make if I put 0.25 pound on each sandwich.

Chapter 13

13.1–13.2 Quiz

1. Terms: $10x, 5, 3x, 1;$
Like terms: $10x$ and $3x$, 5 and 1

2. Terms: $-2n, 7n, -r, 10r;$
Like terms: $-2n$ and $7n$, $-r$ and $10r$

3. Terms: $-12h^2, -4, 9, -3h^2;$
Like terms: $-12h^2$ and $-3h^2$, -4 and 9

4. Terms: $1.4c, 11.4, -2c, -7.3c;$
Like terms: $1.4c$, $-2c$, and $-7.3c$

5. $-7v$ **6.** $3d + 5$ **7.** $9x + 5$

8. $8x - 12$ **9.** $-x - 4$ **10.** $-5.6k - 7$

11. $6m + 4$ **12.** $-4c + 7\frac{2}{3}$ **13.** $\frac{1}{2}(d + 12)$

14. $-3.6(z + 3)$ **15.** $16m$

16. $(3c + 12)$ ft^2 **17.** $\$(0.5g + 0.5)$

13.3–13.5 Quiz

1. $b = -5$ **2.** $q = -7$ **3.** $x = -7\frac{1}{6}$

4. $w = 6.2$ **5.** $n = -35$ **6.** $p = -\frac{2}{11}$

7. $z = -2$ **8.** $a = 5$

9. $h - 3 = -12; h = -9$

10. $b + 7.5 = 4.8; b = -2.7$

11. $-\frac{3}{4}x = -27; x = 36$

12. $\frac{m}{-3.2} = 15; m = -48$

13. $x + 6.3 + 5.6 + 8.4 = 27.4; x = 7.1$ cm

14. 88 ft **15.** $\frac{7}{8}b = 35$; \$5

16. 2.5 mi

Test A

1. Terms: $7j, 3, -4j, 9;$
Like terms: $7j$ and $-4j$, 3 and 9

2. Terms: $10a, 4b, 3a, -15b;$
Like Terms: $10a$ and $3a$, $4b$ and $-15b$

3. Terms: $x^2, -3x, 4, 2x^2, -x, -12;$
Like Terms: $-3x$ and $-x$, 4 and -12

4. $3y + 5$ **5.** $8p - 13$ **6.** $4h - 7$

Copyright © Big Ideas Learning, LLC
All rights reserved.

Answers for Assessment Book

7. $11b - 33$ **8.** $9(x - 4)$ **9.** $\frac{1}{5}(k + 75)$

10. $11x + 12$ **11.** $(40x + 150)$ in.^{2}

12. $6p + 2$ **13.** $n = 27$ **14.** $d = -13$

15. $x = 8.6$ **16.** $z = -16$ **17.** $y = -5$

18. $f = -2.4$

19. $q + 11 = -15; q = -26$

20. $m - 30 = 10; m = 40$

21. $\frac{1}{3}t = 7; t = 21$

22. $\frac{5 + d}{-2} = 14; d = -33$

23. $-4 + t = 8$; 12°F

24. $s + 15 = 40$; \$25

25. $4n + 2.5n = 39$; 6 packs

26. $96 + 12c = 156$; 5 cartons

Test B

1. Terms: m, $3n$, $6m$; Like terms: m and $6m$

2. Terms: $-2t$, $3t^2$, -10, $-14t$;
Like terms: $-2t$ and $-14t$

3. Terms: $5x^2$, $7x$, -3, $10x^2$, $-5x$, -20;
Like terms: $5x^2$ and $10x^2$, $7x$ and $-5x$, -3 and -20

4. 5 **5.** $-y - 1$ **6.** $24a - 29$

7. $4w + 3$ **8.** $2.6(x + 9)$ **9.** $-\frac{3}{5}\left(k + \frac{1}{2}\right)$

10. $11x + 8$ **11.** $4x + 16$ **12.** $4n$

13. $y = -46.35$ **14.** $m = \frac{3}{8}$ **15.** $x = -13.02$

16. $v = 23$ **17.** $x = -5$ **18.** $x = -9$

19. $x = 11, x = -11$ **20.** $x = 5, x = -9$

21. $a + (-12) = 6; a = 18$

22. $45 = \frac{n}{3}; n = 135$

23. $2.1 - 2p = -4.7; p = 3.4$

24. $\frac{1}{2}\left(z + 3\frac{1}{2}\right) = 4\frac{1}{2}; z = 5\frac{1}{2}$

25. $8g + 20 = 76; g = 7$

26. 5 years **27.** 5 volunteers

Alternative Assessment

1. a.

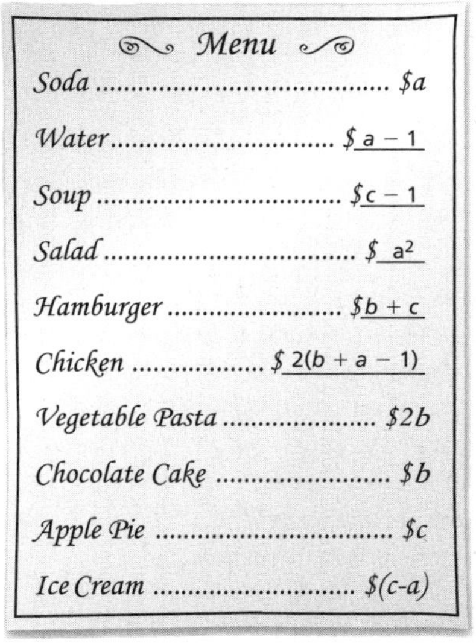

b. $C = a^2 + 2b + 2c$

c. \$18

d. \$22.14

2. a.
$$3x + \frac{2}{3} = 7\frac{5}{6}$$
$$3x = 7\frac{5}{6} - \frac{2}{3}$$
$$3x = \frac{43}{6}$$
$$x = \frac{43}{6} \div 3$$
$$= \frac{43}{6} \times \frac{1}{3}$$
$$= \frac{43}{18}$$
$$= 2\frac{7}{18}$$

Copyright © Big Ideas Learning, LLC
All rights reserved.

b.

$$3x + \frac{2}{3} = 7\frac{5}{6}$$

$$\frac{3x}{3} + \left(\frac{2}{3} \div 3\right) = \left(7\frac{5}{6} \div 3\right)$$

$$x + \left(\frac{2}{3} \times \frac{1}{3}\right) = \left(\frac{47}{6} \times \frac{1}{3}\right)$$

$$x + \frac{2}{9} = \frac{47}{18}$$

$$x = \frac{47}{18} - \frac{2}{9}$$

$$= \frac{47}{18} - \frac{4}{18}$$

$$= \frac{43}{18}$$

$$= 2\frac{7}{18}$$

c. *Sample answer:* Both approaches require several steps before the equation is solved. In both of them it is necessary to use the reciprocal of 3 to divide a fraction by 3. With both methods, the same answer is obtained, but that is what you would expect. There are more steps and more complicated calculations to do when you divide both sides by 3 first, so there are more chances to make an error. For an equation like this one, it is more efficient to subtract first to isolate the expression with the variable.

Chapter 14

14.1–14.3 Quiz

1. $\frac{12}{5}$

2. $\frac{1}{4}$

3. 15.5 miles per gallon

4. $1.20 per box

5. no

6. yes

7. yes

8. no

9. $\frac{12}{3} = \frac{16}{p}$

10. $\frac{e}{8} = \frac{30}{10}$

11. $x = 8$

12. $p = 72$

13. 8 pictures per minute

14. no; On Monday you swim one lap in 2.5 minutes and on Tuesday you swim one lap in 3.0 minutes.

15. Add 6 ounces of Chemical B.

16. $\frac{5}{3} = \frac{25}{c}$; $c = 15$

14.4–14.6 Quiz

1. $b = 42$

2. $k = 4.5$

3. $m = 10.8$

4. $\frac{1}{6}$

5. 1

6.

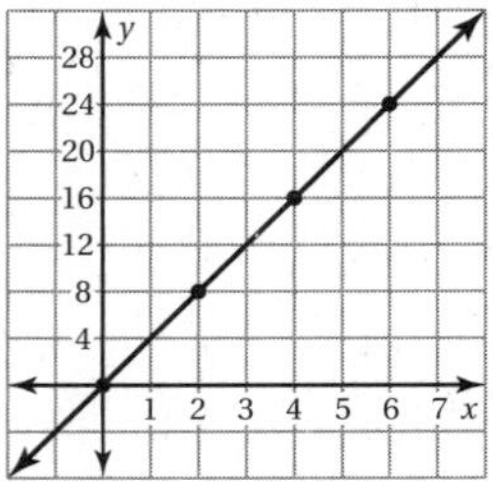

The slope is 4. So, there are 4 inches for each day.

7.

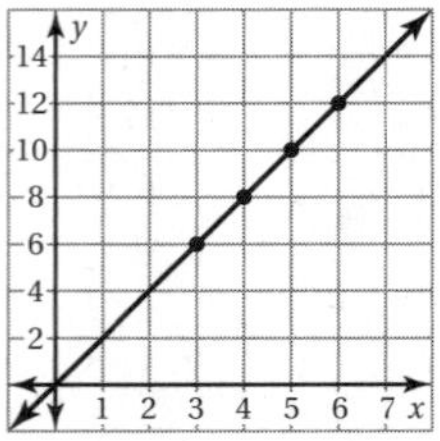

The slope is 2. So, the cost is $2 per pound.

8. no; The equation cannot be written as $y = kx$.

9. yes; The equation can be written as $y = kx$.

10. 15 girls, 10 boys

11. **a.** You: 50; You are driving 50 miles per hour. Friend: 30; Your friend is driving 30 miles per hour.

b. 20 miles per hour

Test A

1. $\frac{5 \text{ girls}}{2 \text{ boys}}$

2. $\frac{8 \text{ players}}{1 \text{ team}}$

3. 81 rotations per min

4. 6 oz per serving

5. yes

6. no

7. yes; Each text message costs $0.12.

8.

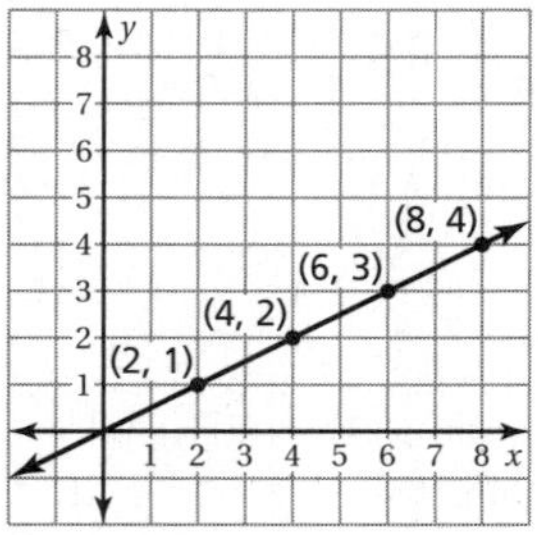

yes

Copyright © Big Ideas Learning, LLC
All rights reserved.

Answers for Assessment Book

9. 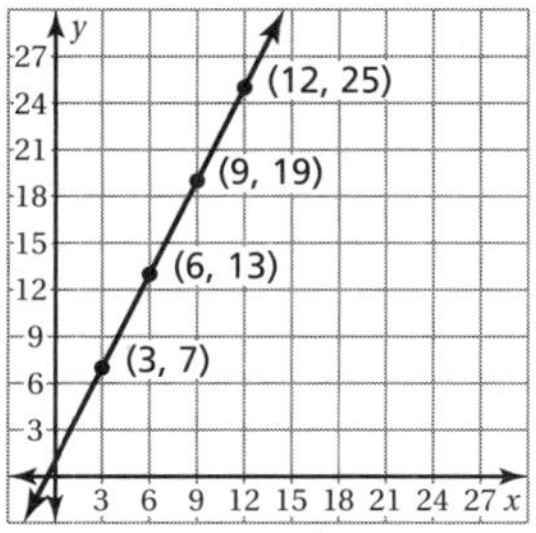

no

10. $\frac{2}{310} = \frac{5}{c}$ **11.** $\frac{m}{3} = \frac{32}{4}$ **12.** $\frac{5}{3} = \frac{12}{c}$

13. $n = 8$ **14.** $p = 308$

15. $k = 5$ **16.** $q = 4.5$

17. 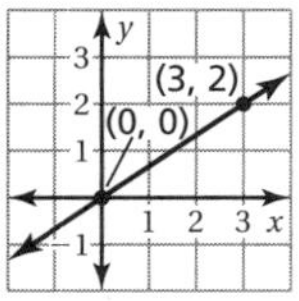

slope: $\frac{2}{3}$

18. 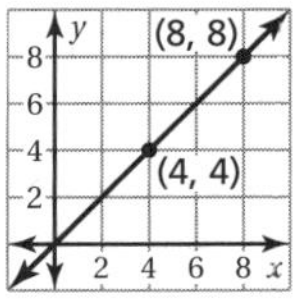

slope: 1

19. yes; The line passes through the origin.

20. no; The equation cannot be written as $y = kx$.

21. no; The equation cannot be written as $y = kx$.

22. 2 games; If g is the number of games the team must win, then $\frac{6 + g}{4} = \frac{2}{1}$; $6 + g = 4(2)$, or 8, so $g = 2$.

23. \$116

24. **a.** skiing: 8
hiking: 6
b. 2 calories
c. 640 calories

Test B

1. $\frac{\text{4 worksheets}}{\text{1 student}}$ **2.** $\frac{\text{5 frogs}}{\text{3 lizards}}$

3. 96 kilometers per hour **4.** \$3.75 per quart

5. no **6.** yes **7.** no **8.** yes

9. no; The costs are \$0.89 per pound, \$0.72 per pound, and \$0.64 per pound to ship packages that weigh 4 pounds, 6 pounds, and 8 pounds, respectively.

10. $\frac{17}{s} = \frac{68}{1000}$ **11.** $\frac{6}{10} = \frac{f}{55}$

12. $\frac{9.95}{5} = \frac{c}{4}$

13. $w = 60$ **14.** $a = 23$ **15.** $d = 9.375$

16. $t = 6$ **17.** $x = \frac{8}{9}$ **18.** $n = 6$

19. 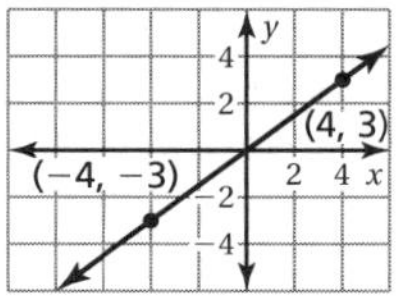

slope: $\frac{3}{4}$

20. 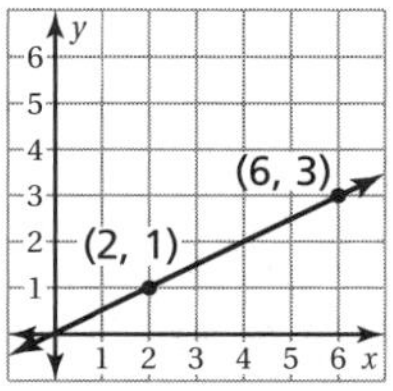

slope: $\frac{1}{2}$

21. yes; The equation can be written as $y = kx$.

22. no; The equation cannot be written as $y = kx$.

23. $k = \frac{2}{3}$; $y = \frac{2}{3}x$ **24.** $k = \frac{1}{5}$; $y = \frac{1}{5}x$

25. $y = 1.61x$

26. 3 cups of milk and 4 cups of flour

27. **a.** You do. **b.** \$127.50

28. $y = 29$; $\frac{y - 4}{6 - 1} = 5$; $\frac{y - 4}{5} = 5$; $y - 4 = 25$; $y = 29$

29. 75 booklets

Copyright © Big Ideas Learning, LLC
All rights reserved.

Answers for Assessment Book

Alternative Assessment

1. a.

	Last Gasp	Greener Gadgets	Tech Things
1 DVD	\$12	\$8.95	\$7.95
3 DVDs	\$10	\$7.95	\$7.95
5 DVDs	\$8	\$6.95	\$7.95
10 DVDs	\$6	\$5.95	\$7.95

b. Last Gasp: \$72
Greener Gadgets: \$62.55
Tech Things: \$71.55
Greener Gadgets has the lowest price.

c. Last Gasp: \$90
Greener Gadgets: \$89.37
Tech Things: \$71.55
Tech Things has the lowest total cost.

2. a. Terry won 6 rounds, Morgan won 4 rounds, and there were 2 ties.

b. Morgan will win the last round because the only card he has left is $\frac{10}{12}$ and the only card Terry has left is $\frac{8}{12}$. Terry will win the game because he won 6 rounds, Morgan won 5 rounds, and there were 2 ties.

Chapter 15

15.1–15.4 Quiz

1. 0.73 **2.** 0.323 **3.** 1.216

4. 65% **5.** 135% **6.** 0.3%

7. 74%, $\frac{3}{4}$, 0.76 **8.** $2\frac{3}{5}$, 262%, 26.2

9. Ben **10.** $\frac{a}{65} = \frac{80}{100}$; 52

11. $\frac{32}{200} = \frac{p}{100}$; 16%

12. $125 = p \bullet 200$; 62.5%

13. $45 = 0.9 \bullet w$; 50

14. 20 questions **15.** 80 athletes

15.5–15.7 Quiz

1. increase; 50% **2.** decrease; 20%

3. decrease; 12.5% **4.** increase; 200%

5. 7.5% **6.** \$52 **7.** 75% **8.** \$18

9. \$60.75 **10.** \$652.50 **11.** 1 year **12.** 5%

13. \$750 **14.** Store A

15. no interest option **16.** \$31.75

Test A

1. 0.667 **2.** 0.001

3. reading email, doing chores, doing homework

4. number of students who walk to school

5. number who play on a sports team

6. greater than

7. $\frac{21}{w} = \frac{35}{100}$; 60 **8.** $\frac{70}{56} = \frac{p}{100}$; 125%

9. $17 = p \bullet 68$; 25% **10.** $a = 0.16 \bullet 80$; 12.8

11. increase; 40% **12.** decrease; 60%

13. \$94.08 **14.** 615 brushes

15. \$73.80 **16.** 35% **17.** \$50 **18.** \$37.12

19. 2 years **20.** \$150 **21.** \$720 **22.** 6.5%

23. \$260 **24.** \$2090 **25.** \$5.88

26. 20% increase

27. Store B; The cost at Store A is \$90.30, at Store B is \$87.75, and at Store C is \$90. So, the cost is the lowest at Store B.

28. 3 years

Test B

1. 0.0005 **2.** 1.28

3. final exam, homework, project, quizzes

4. 6th-graders who participate in sports

5. 7th-graders

6. Write the portions in the same form: 0.18 play sports and 0.45 participate in other activities. $2 \times 0.18 = 0.36$ and $0.36 < 0.45$ so yes, it is true. (2.5 times as many 7th-graders participate in other activities as play sports)

Copyright © Big Ideas Learning, LLC
All rights reserved.

7. $\frac{a}{60} = \frac{45}{100}$; 27 **8.** $\frac{48}{160} = \frac{p}{100}$; 30%

9. $8 = 0.005 \bullet w$; 1600 **10.** $182 = p \bullet 130$; 140%

11. increase; 100% **12.** decrease; 0.8%

13. 2660 miles **14.** 49 degrees **15.** 80%

16. $295 **17.** $20.09 **18.** $362.50

19. $104 **20.** $250 **21.** 1.5 years

22. 4.5% **23.** $1452.50 **24.** $5225

25. 20 gal

26. a. 50% increase **b.** $18

27. $54.34

28. 20 years; The interest earned after t years is $500(0.05)t$, so the balance after t years is $500 + 500(0.05)t$. When the investment doubles, $500 + 500(0.05)t = 1000$; $500(0.05)t = 500$; $0.05t = 1$; $t = 1 \div 0.05 = 20$; 20 years

Alternative Assessment

1. a. *Sample answer:* Top Down Values because it has the greatest first week markdown.

b. Top Down Values

Week 1: $42

Week 2: $33.60

Week 3: $30.24

Base Minus Sales

Week 1: $48

Week 2: $38.40

Week 3: $30.72

Bargains All The Time

Week 1: $54

Week 2: $43.20

Week 3: $30.24

c. *Sample answers:* The prices drop throughout the weeks of the sales. The markdown percents during the second week are the same at all three stores. Base Minus Sales has the same percent markdown all three weeks. The markdowns at the other two stores are the same three markdowns, but in different orders.

d. week 1: Top Down Values; week 2: Top Down Values; week 3: Top Down Values and Bargains All The Time.

e. Top Down Values and Bargains All The Time have the same week-3 prices because of the Commutative Property of Multiplication:

$$60 \times 0.7 \times 0.8 \times 0.9 = 60 \times 0.9 \times 0.8 \times 0.7$$

2. Both Martha and Abigail had the same year 1 balance, so they both earned 3% of $500 the first year. Their balances were different by a small amount at the end of year 2, so they were not both earning 3% of $500 that year. If you find 3% of $515, you get $15.45 and if you add that to $515, you get $530.45, which is Abigail's year 2 balance. If you continue to find 3% interest on the "new" balance instead of on $500 and add it to the "new" balance each year, at the end of 10 years the balance would be $671.96. Abigail takes the money out of her account every year and starts a new account for that year using the principal and the interest as the new principal.

End-of-Course Test 1

1. $4x - 12$; Distributive Property

2. $21x$; Commutative Property of Multiplication and Associative Property of Multiplication

3. 12 units^2 **4.** $4a + 2c$; $14

5. $\frac{14}{15}$ **6.** 6 **7.** 0.42

8. 29.52 **9.** $14\frac{11}{15}$ in.^{2} **10.** 9

11. 17 **12.** 0 **13.** 8

14. 2 c of flour, $\frac{2}{3}$ c of sugar, $1\frac{1}{3}$ c of chocolate chips

15. 20 units^2

16.

Stem	Leaf
3	4 5 9
4	1 2 5 5 7 8
5	6 8
6	2

Key: 4 | 7 = 47

17. −4, −2, 3, 5, 7 **18.** −5, −3, −1, 0, 3

19. the eighteen-pack ($0.30 per juice box compared to $0.35 per juice box in the twelve-pack)

20. 37.5% **21.** 120% **22.** 325%

Copyright © Big Ideas Learning, LLC
All rights reserved.

Answers for Assessment Book

23. They are all in agreement.

$$14.6\% = \frac{292}{2000} = 0.146$$

24. 6 vertices **25.** \$17.50

26. 4 : 6; For every 4 basketballs, there are 6 footballs.

27. about 4 hours, 20 minutes

28. mean: 7; median: 6; mode: 6; IQR: 4; range: 9

29. no; You also need to know how many she has shot so far. There is not enough information to answer the question.

30–33.

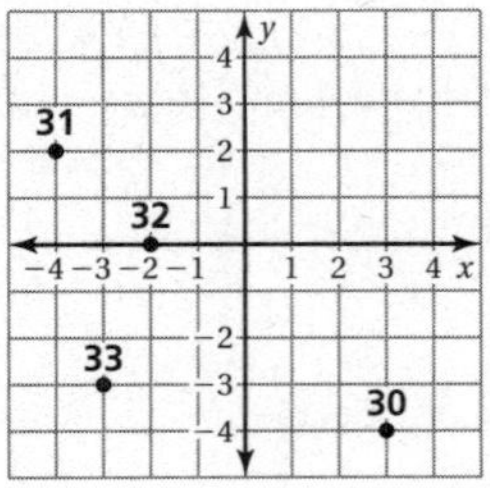

34. $s = 10$ **35.** $c = 6$ **36.** 1500 ft^2

37. $(4 \bullet 6)x = 96$; 4 cm; 128 cm^2

38. 7.125 **39.** $t < 7$ **40.** $m \geq -3$

41. statistical; You expect variability in the answer.

42. not statistical; There is only one answer.

43. 50% of the time

44. Store A: symmetric; Store B: skewed-right

45. Store B **46.** 15

47. 6 **48.** 28 days **49.** 7

50.

Teachers	2	3	10
Students	24	36	120

51.

Cups	52	36	20
Quarts	13	9	5

52. 0–1 **53.** 20 students **54.** 70%

55. $-4, -|2|, -1, |-3|, 4$ **56.** $-7.5, -\frac{36}{5}, 9.5, \frac{21}{2}$

57. 7 **58.** 13 **59.** −12 **60.** −9

61. $\frac{5}{12}$ **62.** −1.36 **63.** $-3\frac{7}{15}$ **64.** −7.5

65. 39 points **66.** $x = \frac{11}{30}$

67. $a = -6$ **68.** 9 pens

69. 1500 ft^2

70. 0.5 minute, or 30 seconds

71. no **72.** yes **73.** $x = 15$

74. $\frac{7}{8}$ **75.** $\frac{1}{2}$ **76.** 50

77. no; Because the additional 15% is taken off the sale price rather than the original price, the discount is actually less than 75% off the original price.

78. \$1344

End-of-Course Test 2

1. $4x + 12$; Distributive Property

2. $14x$; Commutative Property of Multiplication and Associative Property of Multiplication

3. 27 square units **4.** $3.5a + 2c$; \$13

5. $\frac{2}{7}$ **6.** $36\frac{4}{5}$ **7.** 1.44

8. 4.6 **9.** $27\frac{1}{5}$ in.^{2} **10.** 9

11. 37 **12.** 18 **13.** 7

14. $2\frac{1}{3}$ c of flour, $\frac{2}{3}$ c of sugar, $1\frac{5}{6}$ c of chocolate chips

15. 20 units^2

16.

Stem	**Leaf**
3	8
4	2 6
5	0 2 5 6 6 8
6	1 6 9

Key: 4 | 2 = 42

17. −5, −3, −2, 6, 9 **18.** −8, −2, −1, 2, 4

19. the eighteen-pack (\$0.305 per juice box compared to \$0.325 per juice box in the twelve-pack)

Copyright © Big Ideas Learning, LLC
All rights reserved.

Answers for Assessment Book

20. 62.5% **21.** 420% **22.** 212%

23. They are all in agreement.

$46.2\% = \frac{231}{500} = 0.462$

24. 5 vertices **25.** $17.85

26. 8 : 6; For every 8 baseballs, there are 6 soccer balls.

27. 5 hours, 12 minutes

28. mean: 9.9; median: 9.5; mode: 9; IQR: 2; range: 4

29. no; You also need to know how many she has shot so far. There is not enough information to answer the question.

30–33.

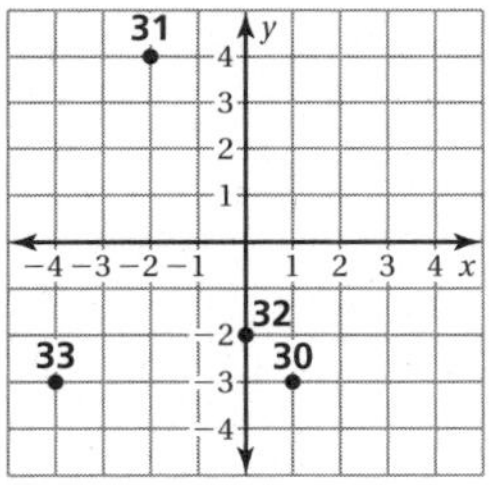

34. $s = 16$ **35.** $c = 8$ **36.** 1456 ft^2

37. $(5 \bullet 6)x = 80$; $2\frac{2}{3}$ cm; $118\frac{2}{3}$ cm^2

38. 3.75 **39.** $w < 5.5$ **40.** $m \geq 7$

41. statistical; You expect variability in the answer.

42. not statistical; There is only one answer.

43. 25% of the time

44. Store A: skewed-left; Store B: symmetrical

45. Store A **46.** 10 **47.** 21

48. 30 days **49.** 10

50.

Boys	3	9	18
Girls	7	21	42

51.

Feet	72	57	39
Yards	24	19	13

52. 8–9 **53.** 20 students **54.** 55%

End-of-Course Test 2

55. $-5, -4, -|-1|, 0, |-2|$ **56.** $-6.7, -\frac{28}{5}, 8.2, \frac{17}{2}$

57. 7 **58.** 5 **59.** −15 **60.** −15

61. $\frac{3}{10}$ **62.** −1.02 **63.** $-2\frac{11}{20}$ **64.** −4.6

65. 23.5 points **66.** $x = -1\frac{7}{30}$

67. $a = -3.2$ **68.** 8 pens

69. 1456 ft^2

70. 0.25 minute, or 15 seconds

71. no **72.** yes **73.** $x = 29\frac{1}{3}$

74. $\frac{2}{3}$ **75.** 2 **76.** 42

77. no; Because the additional 15% is taken off the sale price rather than the original price, the discount is actually less than 90% off the original price.

78. $1325.50

Copyright © Big Ideas Learning, LLC
All rights reserved.